PHYSICS THROUGH THE 1990s

Condensed-Matter Physics

Panel on Condensed-Matter Physics
Physics Survey Committee
Board on Physics and Astronomy
Commission on Physical Sciences,
Mathematics, and Resources
National Research Council

NATIONAL ACADEMY PRESS
Washington, D.C.   1986

NATIONAL ACADEMY PRESS 2101 Constitution Avenue, NW Washington, DC 20418

NOTICE: The project that is the subject of this report was approved by the Governing Board of the National Research Council, whose members are drawn from the councils of the National Academy of Sciences, the National Academy of Engineering, and the Institute of Medicine. The members of the committee responsible for the report were chosen for their special competences and with regard for appropriate balance.

This report has been reviewed by a group other than the authors according to procedures approved by a Report Review Committee consisting of members of the National Academy of Sciences, the National Academy of Engineering, and the Institute of Medicine.

The National Research Council was established by the National Academy of Sciences in 1916 to associate the broad community of science and technology with the Academy's purposes of furthering knowledge and of advising the federal government. The Council operates in accordance with general policies determined by the Academy under the authority of its congressional charter of 1863, which establishes the Academy as a private, nonprofit, self- governing membership corporation. The Council has become the principal operating agency of both the National Academy of Sciences and the National Academy of Engineering in the conduct of their services to the government, the public, and the scientific and engineering communities. It is administered jointly by both Academies and the Institute of Medicine. The National Academy of Engineering and the Institute of Medicine were established in 1964 and 1970, respectively, under the charter of the National Academy of Sciences.

The Board on Physics and Astronomy is pleased to acknowledge generous support for the Physics Survey from the Department of Energy, the National Science Foundation, the Department of Defense, the National Aeronautics and Space Administration, the Department of Commerce, the American Physical Society, Coherent (Laser Products Division), General Electric Company, General Motors Foundation, and International Business Machines Corporation.

Library of Congress Cataloging-in-Publication Data
Main entry under title:

Condensed-matter physics.

(Physics through the 1990s)
Bibliography: p.
Includes index.
1. Condensed matter. I. National Research Council
(U.S.). Panel on Condensed-Matter Physics. II. Series.
QC173.4.C65C66   1985   530.4   85-21778
ISBN 0-309-03577-5

Printed in the United States of America

First Printing, April 1986
Second Printing, October 1986
Third Printing, March 1987

PANEL ON CONDENSED-MATTER PHYSICS

ALEXEI A. MARADUDIN, University of California, Irvine, *Chairman*
NEIL W. ASHCROFT, Cornell University
JOHN D. AXE, Brookhaven National Laboratory
PRAVEEN CHAUDHARI, IBM T.J. Watson Research Center
C. PETER FLYNN, University of Illinois
JERRY GOLLUB, Haverford College
BERTRAND I. HALPERIN, Harvard University
DAVID L. HUBER, University of Wisconsin
RICHARD M. MARTIN, Xerox Corporation
DOUGLAS L. MILLS, University of California, Irvine
ROBERT C. RICHARDSON, Cornell University
JOHN M. ROWELL, AT&T Bell Laboratories

iii

iv

Preface

In this survey of condensed-matter physics we describe the current status of the field, present some of the significant discoveries and developments in it since the early 1970s, and indicate some areas in which we expect that important discoveries will be made in the next decade. We also describe the resources that will be required to produce these discoveries.

Condensed-matter physics is divided roughly into two broad subareas devoted, respectively, to solids and to liquids. In this volume the subarea of solids is subdivided into several subfields, including the electronic properties of solids, their structures and vibrational excitations, critical phenomena and phase transitions, magnetic properties of solids, semiconductors, defects and diffusion, and surfaces and interfaces. The subarea of liquids is divided into the subfields of classical liquids, liquid crystals, polymers and nonlinear dynamics instabilities, and chaos. The subareas of solids and liquids are roughly linked by the subfield of low-temperature physics, which is concerned with phenomena occurring in both of them. This subdivision of condensed-matter physics reflects the manner in which the community organizes itself, through its conferences, workshops, and seminars.

Each of the subfields was reviewed by a member of the community working in that subfield, chosen both for technical expertise and scientific breadth who, in general, had the assistance of many other members of that community. These reviews of the subfields of condensed-matter physics were supplemented by reviews of the new materials that are exciting interest because of the unusual physical properties that they display and the opportunities for technological applications that they may afford, of new experimental techniques whose use has led to remarkable discoveries, and of the National Facilities that have provided researchers in condensed-matter physics with capabilities beyond those available in their own institutions. These reviews were also prepared by experts in the corresponding subject areas.

This volume is organized as follows. Part I is devoted to a discussion of the importance of condensed-matter physics; to brief descriptions of several of the most significant discoveries and advances in condensed-matter physics made in the 1970s and early 1980s, and of areas that appear to provide particularly exciting research opportunities in the next decade; and to a presentation of the support needs of condensed-matter physicists in the next decade and of recommendations aimed at their provision. In Part II, the subfields of condensed-matter physics are reviewed in detail. The volume concludes with several appendixes in which new materials, new experimental techniques, and the National Facilities are reviewed.

As one reads through this volume, one cannot help being struck with the conclusion that condensed-matter physics is an intellectually exciting field of physics in which discoveries have had, and are continuing to have, significant impacts on other fields of physics, as well as on chemistry, mathematics, and the biological sciences. At the same time, it is the field of physics that has the greatest impact on our daily lives through the technological developments to which it gives rise. It has witnessed a decade in which remarkable discoveries and advances in our understanding of the condensed states of matter have been made. It is currently experiencing a period of intensive activity in existing subfields and growth of new subfields, and it offers the promise of significant new discoveries and advances in the decade to come. However, research in condensed-matter physics at a world-class level today is becoming increasingly sophisticated in both theoretical and experimental techniques. With this increasing sophistication is associated a rapidly increasing cost of doing research, in dollars and in manpower, which must somehow be met if the opportunities facing this field are to be achieved. This is a challenge that together with the opportunities will be facing condensed-matter physics in the United States in the next decade.

Finally, I am grateful for the technical contributions of the members of the Panel on Condensed-Matter Physics and for their assistance in drafting the recommendations made in this report. In addition, I want to thank the many members of the U.S. condensed-matter physics community who contributed to every part of this survey, either by writing parts of it or by reading it and making suggestions for its improvement. They are listed at the end of this volume. Their valuable contributions are greatly appreciated.

Contents

II A DECADE OF CONDENSED-MATTER PHYSICS

The Interface Between Solids and Dense
   Media, 157
Theory, 159
Opportunities, 160

APPENDIXES

I

Highlights, Opportunities, and Needs

The focus in this volume is solely on condensed-matter physics, which is the foundation of a significant portion of the broader field of materials science, and the dividing line between the two fields is not always a sharp one. However, we are *not* surveying materials science nor the considerable impact of condensed-matter physics on technology. The interface between physics and technology will receive fuller treatment in another volume in this survey.

CONDENSED-MATTER PHYSICS AND ITS IMPORTANCE

Condensed-matter physics is the fundamental science of solids and liquids, states of matter in which the constituent atoms are sufficiently close together that each atom interacts simultaneously with many neighbors. It also deals with states intermediate between solid and liquid (e.g., liquid crystals, glasses, and gels), with dense gases and plasmas, and with special quantum states (superfluids) that exist only at low temperatures. All these states constitute what are called the *condensed states* of matter.

Condensed-matter physics is important for two reasons. The first is that it provides the quantum-mechanical foundation of the classical sciences of mechanics, hydrodynamics, thermodynamics, electronics, optics, metallurgy, and solid-state chemistry. The second is the massive contributions that it provides to high technology. It has been the source of such extraordinary technological innovations as the transistor, superconducting magnets, solid-state lasers, and highly sensitive detectors of radiant energy. It thereby directly affects the technologies by which people communicate, compute, and use energy and has had a profound impact on nonnuclear military technology.

At the fundamental level, research in condensed-matter physics is driven by the desire to understand both the manner in which the building blocks of condensed matter—electrons and nuclei, atoms and molecules—combine coherently in enormous numbers ($\sim 10^{24}$/cm^3) to form the world that is visible to the naked eye, and much of the world that is not, and the properties of the systems thus formed. It is in the fact that condensed-matter physics is the physics of systems with an enormous number of degrees of freedom that the intellectual challenges that it presents are found. A high degree of creativity is required to find conceptually, mathematically, and experimentally tractable ways of extracting the essential features of such systems, where exact treatment is an impossible task.

Condensed-matter physics is intellectually stimulating also because of the discoveries of fundamentally new phenomena and states of matter, the development of new concepts, and the opening up of new subfields that have occurred continuously throughout its 60-year history. It is the field in which advances in quantum and other theories most directly confront experiment and has repeatedly served as a source or testing ground for new conceptual ways of viewing complex systems. In fact, condensed-matter physics is unique among the various subfields of physics in the frequency with which it feeds its fundamental ideas into other areas of science. Thus, advances in such subareas of condensed-matter physics as many-body problems, critical phenomena, broken symmetry, and defects have had a major impact on nuclear physics, elementary-particle physics, astrophysics, molecular physics, and chemistry. These advances continue and offer the promise of equally fundamental discoveries in the next decade.

At the same time, condensed-matter physics excites interest because of the well-founded expectations for applications of discoveries in it. Of all the branches of physics, condensed matter has the greatest impact on our daily lives through the technological developments to which it gives rise. Such familiar devices as the transistor, which has led to the miniaturization of a variety of electronic appliances; the semiconductor chip, which has made possible all the myriad aspects of the computer; magnetic tapes used in recording of all kinds; plastics for everything from kitchen utensils to automobile bodies; catalytic converters to reduce automobile emissions; composite materials used in fan jets and modern tennis rackets; and NMR tomography are but a few of the practical consequences of research in condensed-matter physics. A whole new technology, optical communications, is being developed at this time from research in condensed-matter physics, optics, and the chemistry of optical fibers.

These examples serve to illustrate the intimate connection between fundamental science and the development of basic new technology in condensed-matter physics. In both universities and industry they are carried out by people with the same research training, who use the same physics concepts and the same advanced instrumentation. Because fundamental science in condensed-matter physics is so deeply involved with technological innovation, it has a strong natural bond with industry. This is the main reason why condensed-matter physics has been so successful in leading industrial innovation.

Indeed, the full extent to which the consequences of research in condensed-matter physics play a role in the quality of our everyday lives, and in meeting national needs, is far greater than any such listing can indicate. In order to show this explicitly we have constructed the matrix displayed in Table 1, the first column of which lists the subareas of condensed-matter physics, and the first row the major areas of human and technological activity that are of national interest. The elements of the matrix are filled in with a solid circle, indicating a critical connection between the corresponding subarea of condensed-matter physics and the area of application; a half-filled circle, indicating an important or emerging connection; an open circle, denoting the possibility of a connection; or a blank, implying that the connection is not known. In Appendix A this matrix is repeated, but with qualitative

TABLE 1 Connections Between Subareas of Condensed-Matter Physics and Applications of National Interest[a]

Subarea	Information Processing	Speech and Data Communications	Energy	Medical	Transportation	Space Technology	National Security
Electronic properties	●	◑	●	○	●	○	●
Phonons/electron-phonon interactions	◑	○	◑	◑	○	○	○
Phase transitions	◑	○	◑	○	◑	○	●
Magnetism	●	●	●	◑	●	◑	◑
Semiconductors	●	●	●	◑	●	●	●
Defects/diffusion	●	●	●		●	◑	●
Surfaces/interfaces	●	●	●	◑	●	◑	●
Low-temperature physics	◑	○	●	◑	○	○	○
Liquids	◑	○	◑	○	◑	◑	○
Polymers	●	◑	◑	●	●	◑	◑
Nonlinear dynamics, instabilities, chaos		○		○	○	○	●

[a] A solid circle indicates a critical connection between the subarea of condensed-matter physics and the area of application; a half-filled circle indicates an important or emerging connection; an open circle denotes the possibility of a connection; a blank implies connection is not known.

comments concerning the connections replacing the various circles. This table makes the point graphically that condensed-matter physics plays an indispensable role in the maintenance of the quality of our daily life and in providing for national security.

DISCOVERY

The 1950s saw such achievements as the rapid development of semiconductor technology after the discovery of the transistor; the rise of many-body theory (the application of the methods of quantum field theory to large and complex systems) as a field of theoretical physics, and its crowning achievement, the solution of the 50-year-old problem of superconductivity; the heyday of magnetic resonance methods in physics; and the elucidation of the Fermi surfaces of metals. The 1960s saw the discovery of high critical fields and superconducting magnets, as well as of the Josephson effect and other electron tunneling methods and devices; the construction of the first working lasers and further giant strides in laser physics; the initial explanation of the ancient problem of the resistance minimum by Kondo and the opening up of a whole new physics of similar Fermi-surface effects in metals such as the x-ray edge; the development of pseudopotential and density functional methods, among others, that have made electronic structure calculations almost routine; and the initial development of high-energy probe methods for the study of electronic structure such as ultraviolet photoelectron spectroscopy (UPS) and x-ray photoelectron spectroscopy (XPS).

From time to time, there have been those who have predicted the end of this era of discovery. Remarkably, the subject continues to produce surprises. In what follows we present a selection of some of the most interesting advances in condensed-matter physics that occurred in the 1970s and early 1980s.

Artificially Structured Materials

One area of condensed-matter physics that has progressed remarkably in the past decade is that of artificially structured materials—materials that have been structured either during or after growth to have dimensions or properties that do not occur naturally.

The most important techniques for the creation of such materials are molecular-beam epitaxy (MBE), the molecule-by-molecule deposition of material of the desired composition from a molecular beam, and metallo-organic chemical vapor deposition (MOCVD). These are prime examples of technological breakthroughs, used primarily to make semiconductor lasers and other devices, feeding back to fundamental physics. One can fabricate artificial periodic superlattices consisting of alternating layers of different semiconductors, different metals, or semiconductors and metals, and one can also create artificial, purely two-dimensional electron gases. The latter have unique and important properties, e.g., extremely high electron mobilities, which cannot be provided by metal-oxide-semiconductor (MOS) inversion layers. The new physical phenomena to which the resulting structures have given rise include the quantized Hall effect and the fractionally quantized Hall effect. It has also been possible to grow metallic superlattices in which the electronic mean free path is appreciably longer than the period of the superlattices (the sum of the thicknesses of the two alternating metal layers). It is found that it is possible to induce new lattice structures rather easily in such superlattices. Metal/insulator superlattices are ideal systems for the study of dimensional effects in metals, e.g., the crossover from two- to three-dimensional superconductivity in Nb/Ge superlattices as the Ge thickness is decreased.

The Quantized Hall Effect

Modern technology has made possible unique, purely two-dimensional electron gases (in the sense that only one quantum state is excited in the direction perpendicular to the plane of the gas, so that electronic motion in it is strictly confined to that plane). These systems show exciting properties and are a new laboratory for the study of fundamental physics. The most remarkable property of such systems is undoubtedly the quantized Hall effect. At low temperature and high perpendicular magnetic field, the electron states are split into so-called Landau or cyclotron energy levels. It is found that when the Fermi level is between two such levels one sees an almost perfectly flat plateau or constant value of the Hall conductance, the conductance perpendicular to the electric and magnetic fields, as well as zero parallel conductance. These plateaus are found to be quantized in units of $e^2/h = 1/25,812.8$ ohm^{-1}. The precision of this result, at least one part in a hundred million, has led to improvement in the measurement of this fundamental constant and to a new portable resistance standard. More recently, quantization of the Hall conductance in simple fractions like 1/3, 2/5, and 2/7 of e^2/h has been seen, and an explanation of this effect has been proposed, and widely accepted, that involves a completely new and unexpected ordered state of matter. In this state one proposes that a new type of elementary excitation with fractional electronic charge plays a major role.

Effects of Reduced Dimensionality

For many years condensed-matter theorists studied one- and two-dimensional models of solids because it was often possible to obtain exact results there where the corresponding, physical, three-dimensional models were intractable. The existence of such exact solutions in low-dimensional systems has prompted experimentalists to search, successfully, for physical systems whose physical properties agree well with those of one- and two-dimensional theoretical models. These include quasi-one-dimensional magnetic systems composed of chains of magnetic atoms, separated from each other by nonmagnetic atoms, and quasi-two-dimensional systems realized by layered compounds, such as graphite intercalation compounds, in which atomic layers are widely separated and weakly interacting.

Other examples have arisen either out of technological discoveries or from the synthesis of interesting new materials. The inversion layers used in the quantized Hall effects are an example of reduced dimensionality systems important in technology, an example that has been vital to the physics of disordered systems as well. Another is the development of methods for studying adsorbed layers on surfaces that undergo phase transitions of typically two-dimensional type. A third is the discovery of methods for making freely suspended layers of a liquid crystal one or a few molecules thick.

New materials showing metallic properties in only one or two dimensions have been synthesized, for instance the transition-metal dichalcogenides, which can be cleaved to produce single layers or intercalated with large molecules that separate the layers by large distances, and a number of organic one-dimensional chain metals such as polyacetylene. These various developments have encouraged experimentalists and theorists to think of dimensionality as a new free parameter.

Charge-Density Waves

Among phenomena that are most clearly demonstrated in low-dimensionality systems are charge (or in some cases spin) density waves. A few isolated cases in which the structure of a solid was modulated periodically had been known for decades, but it was not until modern low-dimensionality materials became available, such as the dichalcogenides and trichalcogenides of Nb and other transition metals, and some organic metals, such as polyacetylene and tetra-thiafulvalene-tetracyanoquinodimethane (TTF-TCNQ), that the phenomenon could be studied in general. Theory has predicted for many years that such materials should show density waves especially easily. In such materials the structure contains two periods that may be incommensurate, hence giving an overall nonperiodic structure. A particularly important possibility is the sliding of such an incommensurate wave through the parent lattice, a new phenomenon illustrating in a clean microscopic model the age-old effects of sliding and sticking friction. The materials that display these effects, e.g., $NbSe_3$ and TaS_3, are remarkable quantum systems with the richness of superconductivity and should be excellent for studying various aspects of macroscopic quantum phenomena. Other interesting phenomena relate to defects in these waves, which have strange topological properties, fractional charge per unit area, and, in the case of polyacetylene, strange spin and charge properties. This subject continues to be actively discussed, not only because of its scientific interest but also because of its possible technical interest.

Disorder

It is only within the past decade that physicists have begun to focus on the problems intrinsic to disordered states of matter such as random alloys, glass, and gels. Historically they had dealt with such systems— often effectively—by trying to *average out* the disorder in the most efficient possible way, to produce an "effective medium." Now they have begun to look for intrinsic properties of disordered materials. The most striking of these is localization, the tendency to form quantum states that cannot move except with the help of thermal energy. Experimentally, the study of localization is much clarified by using a two-dimensional geometry, in which one often sees a unique nonclassical behavior of the electronic conductivity, and by technical advances in microfabrication, which allow the study of effectively one-dimensional wires and of tiny loops that show strange conductivity oscillations in a magnetic field. A second disordered material of technical importance is glass; the glass transition and the high-temperature annealing properties of glass remain almost completely mysterious, but a whole new physics has grown up around a new entity recently discovered in the low-temperature behavior of glass, the so-called tunneling centers. The structure of glass is also a great mystery; the computer may help in deciphering it, but in fact we know so little that we do not yet even believe we can program a computer to make a viable model of glass.

Mixed Valence and Heavy Fermions

It is not uncommon, the chemists have found, for the same chemical element to exhibit two valences in the same compound—as, for example, magnetite, which contains both ferrous and ferric iron at different atomic positions. On the other hand, metals such as nickel do not necessarily have a fixed valence, as the electrons move freely through the lattice. The rare-earth metals, however, normally have a fixed valence for the inner f electrons, which can be identified because they show magnetic properties identical to those of ions in an insulating salt. It now appears that there is a large class of compounds based on the rare-earth atoms Ce, Sm, Eu, Tm, Yb, and now the actinide element U, that are intermediate between these two cases in an unusual way. Some types of measurements—one-electron probes, x-ray edges—show both valences simultaneously developed on the same atom. Other types of measurement, such as those of low-temperature magnetism or conductivity, show a fixed valence, sometimes intermediate and sometimes not. It appears that electrons are quantum mechanically tunneling rather slowly in and out of the f shells, with very exotic results, such as electron bands with effective electron masses as large as 1000 times a normal electron mass, which nonetheless exhibit superconductivity at very low temperatures. Present speculation is that these superconductors are of a totally new type, and are analogous to superfluid ^{3}He. The valence fluctuations in other materials lead to a number of other fascinating effects: metal/insulator transitions, magnetic/nonmagnetic transitions, soft (highly compressible) lattices, and transitions into exotic magnetic ground states. A full explanation of these phenomena might have far-reaching consequences for our understanding of magnetism and bonding in solids.

The Superfluid Phases of ^{3}He

A high point in research in condensed-matter physics of the last decade was the discovery that ^{3}He is a superfluid (i.e., can flow without resistance through narrow channels) at temperatures below 3 mK. This is the first, and only, new superfluid to be discovered since the superfluidity of ^{4}He was established in 1937. The properties of ^{3}He are very different from those of ^{4}He because ^{4}He obeys the quantum-mechanical laws of Bose statistics, whereas ^{3}He obeys Fermi statistics, the same as electrons. At the same time, superfluid ^{3}He displays a rich variety of physical properties in addition to those possessed by the previously known superfluids. This is because the interaction between pairs of helium atoms that is responsible for the superfluidity of ^{3}He is qualitatively different from the interaction between pairs of electrons responsible for the superconductivity of all the currently known superconductors. In particular the superfluid is locally anisotropic, acting as though it was made up of molecules with internal rotational motions about a specific direction.

Several major advances in condensed-matter physics were fueled primarily by new theoretical concepts. Descriptions of two of them follow.

The Renormalization Group Methods

These techniques are useful in dealing with physical phenomena in which there exist fluctuations that occur simultaneously over a wide range of different length, energy, or time scales. The method proceeds by stages, in which one successively discards the shortest-wavelength fluctuations until a few macroscopic degrees of freedom remain. The effects of the short-wavelength fluctuations are taken into account approximately at each stage by a *renormalization*, i.e., change in magnitude, of the interactions among the remaining long-wavelength modes. These techniques were developed initially in particle physics but came into their own in the theory of phase transitions, the branch of condensed-matter physics that deals with changes of state, such as the melting and freezing of solids and liquids and the magnetization of ferromagnets. Their use has provided a theoretical understanding of empirical relations among different properties near the phase transition or critical point of a given system and has made it possible to predict critical properties with a high degree of accuracy. These predictions have been confirmed by a wide variety of subsequent experiments. The renormalization group techniques have found applications in such diverse areas of condensed-matter physics as disordered electronic systems, impurity problems, disordered magnetic materials called spin glasses, nonlinear dynamical systems, long polymer chains, and per-colation through macroscopically inhomogeneous systems such as porous rocks.

Chaotic Phenomena in Time and Space

A second new subfield of condensed-matter theory has arisen from our overlapping interests with the fields of hydrodynamics and plasma physics in strongly driven systems, systems so far from equilibrium that linear equations no longer hold. A bewildering variety of phenomena can appear from the simplest models of these systems, which attempt to describe them in terms of one or a few degrees of freedom: singly or doubly periodic motions, or purely chaotic ones, arising entirely from the internal dynamics with no random external influences. Some remarkable universality properties appear in such systems and are exhibited under some experimental conditions. Many focuses of debate remain: for instance, it is clear that fully developed turbulence cannot be described by such simple mathematics, but how far can such a simple approach go toward providing a description? What is the criterion that determines the remarkable patterns that often appear in such systems? We have learned how complex the simplest sets of equations describing a few modes can be—how far do we have to go to describe a realistic system like a real laser or a solidifying liquid? Chaotic nonlinear behavior is common to a wide variety of condensed-matter systems, such as semiconducting lasers and various superconducting devices, as well as the much studied hydrodynamic systems.

Finally, some new areas have resulted from experimental breakthroughs. Some of these are discussed below.

Widespread Use of Synchrotron Radiation

Synchrotron radiation is electromagnetic radiation emitted from particle accelerators by charged particles (usually electrons) with large energy in the range from hundreds of MeV to 10 GeV or more. Synchrotron sources provide intense radiation at wavelengths for which laser sources are either unavailable or not yet tunable. Because synchrotron radiation has a number of desirable characteristics, e.g., high brightness, wide tunability, strong collimation, linear polarization, stability, and the fact that the radiation often occurs in ~0.1-1 nanosecond pulses, in the past 10 years this waste product of particle physics has been used increasingly for low-energy physics in a broad range of fields. In condensed-matter physics, synchrotron radiation has been used to determine experimentally the energy-momentum relation $E(\mathbf{k})$ for electrons in such elements as Cu and Ni and such semiconductors as GaAs and CdS; the inadequacy of a purely band model of ferromagnetism in nickel has been demonstrated by an experimental determination of the temperature dependence of the exchange splitting; an understanding of the fundamental problem of two-dimensional melting and wetting has been gained through experiments employing synchrotron radiation; the structure of glassy amorphous materials has been investigated in this fashion, and phase transitions in few-molecule-thick layers of liquid crystals have been studied; and the oxidation state and local geometries of molecules adsorbed on surfaces and studies of catalytic activity at surfaces and of the structures of the surfaces themselves have been carried out. A great deal of synchrotron radiation work has been carried out on the formation of semiconductor-oxide, semiconductor-semiconductor, and semiconductor-metal inter-faces; synchrotron radiation has also been used in lithography to produce artificial structures with dimensions as small as 70 Å. Synchrotron radiation today stands as one of the most versatile tools available to experimentalists in a broad range of fields.

Atomic Resolution Experimental Probes

A major advance of the past decade in instrumentation for experimental studies of condensed matter was the development of several probes capable of seeing individual atoms. One of these is the scanning vacuum tunneling microscope. In this instrument a sharp metal tip is placed at a distance of ten or so angstroms above a solid surface, with a potential difference between the tip and the surface. Electrons can transfer from one to the other via quantum-mechanical tunneling, and the resulting electric current is sensitive to the distance from the tip to the sample. As the tip is scanned across the surface the height of the surface at each point can be determined from the current fluctuations. In particular, bumps in the surface electron density produced by individual atoms can influence the current, as can steps and other defects. Thus with this instrument one now has the possibility of determining the structure by direct observation. Also developed in the past decade are new electron microscopes that can be used to image defects within the bulk of a crystal. By 1982 the resolution of commercial transmission electron microscopes had reached about 1.5 Å. Since the beam passes through the entire sample, electron microscopy is naturally adapted to the study of linear and planar defects aligned with the beam. Atomic positions in perfect crystals can also be obtained through the use of thin-film samples. A related atomic resolution probe is the scanning transmission electron microscope whose current resolution is about 2 Å. Heavy atoms located on carbon films have been imaged individually by this probe. We can expect that in the future these kinds of probe will be used more and more for direct determination of atomic positions within and on the surfaces of solids.

RESEARCH OPPORTUNITIES IN CONDENSED-MATTER PHYSICS IN THE NEXT DECADE

The list of outstanding achievements during the past decade given in the previous section demonstrates the vitality of condensed-matter physics and is a strong indication of progress to be made in the coming decades. Many of the developments provide new experimental or theoretical tools for studying physical problems that are as yet unsolved. As examples, we cite the applications of vacuum tunneling microscopy and of intense synchrotron radiation sources, as well as the various renormalization-group methods of theoretical analysis. There are newly discovered materials or phenomena that are only partially understood and that therefore open up new fields of inquiry. Examples here include the heavy-fermion superconductors, modulated semiconductor structures, the quantized and fractionally quantized Hall effect, and the random magnetic-field problem.

In the remainder of this section we describe some of the areas of condensed-matter physics that appear to us to provide particularly exciting research opportunities in the next decade. In compiling such a list, we are, however, cognizant of the fact that each of the preceding two or three decades has seen the discovery of physical phenomena or methods that could not have been predicted at the beginning of that decade. For example, several of the outstanding discoveries listed in the previous section, such as the new phases of ^{3}He and the quantized Hall effect, would have been absent from a list of opportunities drawn up in 1970. It is virtually certain that the coming decade will have its share of such unexpected discoveries.

A great deal of effort is expected to be devoted to the determination of the structures and excitations of surfaces of crystalline solids, both clean and covered with adsorbed layers, and of the interfaces between two different solids and between solids and liquids. These investigations will be carried out by the use of such instruments and techniques as the recently developed scanning vacuum tunneling microscope, Rutherford ion backscattering, grazing-incidence x-ray scattering, low-energy electron diffraction, electron energy loss spectroscopy, and atomic diffraction. The results of these determinations are crucial to a detailed understanding of various surface and interface excitations, electronic, vibrational, and magnetic. We can hope that eventually we may not only determine surface structures, but control them by deliberate doping or other physical and chemical techniques. The great goal of understanding catalysis must always remain in our minds, but few experiments yet undertaken approach the real problems of catalysis in practical systems as opposed to atomically clean models.

The determination of the structure, ground-state properties, and elementary excitations in glasses, amorphous materials, and other disordered systems both magnetic and nonmagnetic will be an active area of research, because of both the interesting physical properties displayed by such systems and the technological importance of many of them. There is certain to be a vigorous effort in developing new types of disordered materials for electronic and optical applications.

The interplay of this field with computer science and even more distant fields such as neurobiology is a fascinating new development that is sure to be a major area of activity in the next decade. Here, for the first time, physics is pushing at the theoretical limits of computational complexity, and hence it is not even clear that meaningful simulation of the structure of glass, or even the folding of a random polymer or protein molecule, for example, can be carried out in principle without recourse to completely new ideas in the computer field and new and complex theoretical concepts. We may also assume that new physical and chemical insights will be necessary in this area, especially as we approach problems involving even more complicated materials such as polymer glasses and gels.

Artificially structured materials can be ordered in their structure, as in the case of artificial superlattices, or they can be ordered on a microscopic scale and yet be disordered overall, as in the cases of systems composed of small particles of one material embedded in a matrix of a second. Such materials possess transport, elastic, and optical properties that can differ considerably from those of their constituents in the bulk state, and, perhaps more importantly, these properties can be tuned in desirable ways by varying the constituents and their thicknesses in the case of superlattices or by altering the constituents, their size distributions, and their relative concentrations in the case of mixed media. The length scales involved in artificially structured materials, however, are so small that many of the conventional methods of solid-state physics are no longer applicable in determining their physical properties. In this area interface states, ballistic transport, Kapitza resistance, quantum-well effects, noise, and electromigration and thermomigration are all topics of fundamental interest and will offer research opportunities for the future.

In the area of phase transitions, we understand equilibrium phenomena in principle very well, but the kinetics of phase transitions remains an important field of materials science, especially because of its practical value, as for example under the exotic conditions of laser pulse annealing. Kinetic questions may even underlie the problems of phase transitions in the early universe. The renormalization group will continue to expand beyond its classic uses into more or less exotic domains, as it has already enlightened studies of chaos and disorder.

The last decade has seen the creation of new organic and polymeric materials with striking physical properties, among them metallic electrical conductivity and even superconductivity. Although a theoretical explanation of *some* of the properties of *some* of these materials is beginning to emerge, much remains to be understood, and opportunities for good theoretical work in this field will continue to exist into the foreseeable future.

The underlying physics of exotic new crystalline materials, such as heavy fermion conductors, charge-density wave materials, and high T_c and H_{c2} superconductors such as the Chevrel phases, is expected to be elucidated in the coming decade.

It has been shown repeatedly that as materials are subjected to more extreme physical conditions they display new physical properties. Thus, within the past decade the combination of small size, low temperature, and high magnetic fields made possible the discovery of the fractionally quantized Hall effect; the ability to reach ultralow temperatures made possible the discovery of the superfluid phases of ^{3}He; the use of very high pressures enabled the rare gas xenon to be solidified into a metal; the use of very low pressures (ultrahigh vacuum) has made possible the first efforts to determine structures of surfaces uncontaminated by unwanted foreign atoms or oxide layers. It is expected that the study of condensed matter at ever lower and higher temperatures, at higher magnetic and electric fields, at higher and lower pressures, and at much higher purities will continue into the next decade, with the range of properties of known materials being broadened thereby.

In the field of nonlinear dynamics, instabilities, and chaos, many questions have been raised by the work of the past decade, and efforts to answer them will surely parallel the discovery of new phenomena in the next decade. We have already mentioned some of these, but a particularly important one is how to treat turbulence and instabilities in real systems with many degrees of freedom when we know that the dynamics of systems with only a few degrees of freedom is already incredibly complicated. The effect of sequences of instabilities on heat, mass, and momentum transport in fluids have not been adequately studied and could have practical applications, e.g., in the understanding of lubrication. There is as yet no really fundamental understanding of the mechanisms that control solidification patterns in, for example, snowflakes, quenched alloys, or directionally solidified eutectic mixtures. An important problem for the future is the computer simulation of the onset and growth of turbulence in hydrodynamic systems. Fundamental questions that remain to be answered include: Is chaos a meaningful concept in quantum mechanics? How does one characterize and classify the steady states that are obtained under constant external conditions away from thermodynamic equilibrium?

While our ability to calculate electronic structures in relatively simple situations with relatively weak interelectronic interactions, such as semimetals and semiconductors, has increased enormously, we still have great difficulties with a variety of types of strongly interacting systems. Theories of magnetic metals and the various types of metal-insulator transitions remain controversial. Even more primitive is our understanding of metals where the electron-phonon interaction is strong, such as the charge-density wave materials as well as the strong-coupled superconductors. Many as yet unexplained experimental data exist. The same difficulties probably extend to many surface electronic states as well as to the newer organic and other linear compounds. The mixed valence problem promises to be an even more difficult one to understand.

The methods of pseudopotential and density functional theory work well for chemically simple systems, but one hopes to extend electronic calculations at the same level of rigor and accuracy to chemically sophisticated systems such as organic compounds. Equally sophisticated molecular orbital methods may be within our grasp, but so far their exploitation has rested with more chemically oriented theorists.

From liquid crystals one should expect that more complex mesophases will begin to be of importance in condensed-matter theory and experiment. A start has been made with lyotropic liquid crystals and membrane phase transitions, for example, as well as with the physics of biologically interesting molecules such as proteins and nucleic acid chains. The next decade may be the decade in which the physics of random polymers becomes as interesting as more conventional disordered systems are now.

Applications of femtosecond laser spectroscopy to studies of condensed matter will increase in number and scope over the coming decade. Structural changes, such as melting and structural phase transitions, can now be studied in a time-resolved fashion on the femtosecond scale. The response of crystalline solids to short electrical pulses can now also be studied by femtosecond spectroscopy. Efforts to create even shorter pulses will continue, in parallel with efforts to extend the wavelengths of such short pulses toward the ultraviolet and infrared regions.

Free-electron lasers show great promise as high-power tunable sources. The first free-electron laser operated in the near infrared, and recently lasing action has been achieved in the blue region of the visible spectrum. Two that produce both far-infrared and infrared radiation are just coming into operation. The availability of such instruments will open the door to the experimental study of a wide variety of nonlinear optical phenomena in condensed matter. These include the generation of magnetic and vibrational excitations with wave vectors at arbitrary points of the first Brillouin zone of the corresponding crystals; driving a displacive ferroelectric crystal above its nominal transition temperature to induce the ferroelectric phase transition at this higher temperature; studying the pinning of charge-density waves in one-dimensional metals; and investigating the expected large nonlinear response of two-dimensional plasmons at semiconductor heterojunction interfaces.

It is expected that epithermal neutrons produced by pulsed neutron sources will be used to study excitations in condensed matter whose energies exceed thermal energies and extend up into the electron-volt range. In addition, they will include time-dependent effects such as high-frequency vibrations in solids, particularly those containing hydrogen atoms. It should also prove possible to measure in this way the momentum distribution of light atoms in their ground state. Of particular interest in this regard is the superfluid ^{4}He, for which a zero-momentum condensate fraction is presumed to exist but whose magnitude remains uncertain.

We believe that the next decade will see the increased use of insertion devices (undulators and wigglers) to increase the brightness of synchrotron radiation sources. This development will make it possible, for example, to study the properties of defects in crystals in the 10^{-5}-10^{-6} concentration range, in contrast with the 10^{-3}-10^{-4} concentration range that can be studied at present. The higher resolution in energy and momentum expected from the use of insertion devices will also make it possible to study small samples of materials that are difficult to prepare in the form of large crystals. Inelastic x-ray scattering from the bulk and from surfaces should become a reality, as well as the ability to determine experimentally electronic structures of solids, particularly of systems with small Brillouin zones, such as artificially structured materials and semiconductors with reconstructed surfaces.

NEEDS OF CONDENSED-MATTER PHYSICS IN THE NEXT DECADE

It is obvious that the significant new discoveries in condensed-matter physics described in the previous section will not appear spontaneously. The question immediately arises: What will it take to produce them?

To answer this question it is necessary first to describe briefly the changes that have occurred in the past decade in the directions of research in condensed-matter physics and the way it is done, which differs from the way research is done in other areas of physics.

The nature of research in condensed-matter physics has changed throughout its history to embrace ever-more complex phenomena. Among its earliest triumphs in the post-World War II era was providing the basic understanding of high-technology materials (e.g., semiconductors, magnetic materials, and superconductors). Over the years the sophistication of condensed-matter theory has reached such a level that at least for elementary systems with almost perfect structures, such bulk properties as their cohesive, electronic, dynamic, thermal, optical, magnetic, and transport properties are now well understood.

In recent times, the emphasis in condensed-matter physics has shifted toward materials with novel or special properties; to imperfect systems; toward problems relating to technologically or biologically important materials and structures; to bounded systems and ones with surfaces and interfaces; to those with remarkable states of electronic order; to strongly perturbed rather than equilibrium systems; and to disorder rather than order. As a consequence, with the aid of new instrumentation, new materials fabrication procedures, and the imaginative use of the computer, condensed-matter physics is beginning to provide the basis for understanding the fundamental properties of systems that are still more interesting than the classical ones from a scientific and technological point of view.

Research in condensed-matter physics differs from research in some other areas of physics in a way that is fundamental for determining priorities in support for it: it is carried out almost entirely by individuals or by small groups of researchers. This is indicated in Figure 1, where a histogram is plotted showing the number of papers published in condensed-matter physics in 1982 in three leading journals as a function of the number of authors. In universities the groups often consist of a faculty member and a graduate or postdoctoral student; in industrial and government laboratories they are likely to consist of colleagues on the staffs of these institutions or of a staff member and a

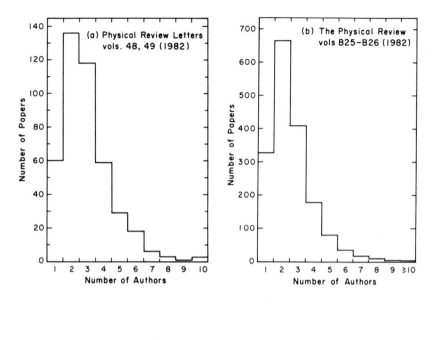

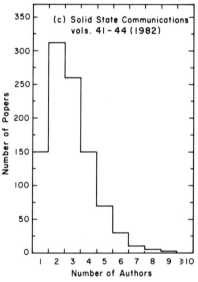

FIGURE 1   Histogram of the number of papers in condensed-matter physics published by leading journals in 1982 as a function of the number of authors: (a) *Physical Review Letters*, (b) *The Physical Review*, (c) *Solid State Communications*.

postdoctoral researcher. It is estimated that there are about 800 such groups throughout the country. This tendency to conduct research in condensed-matter physics individually or in small groups is displayed both by theorists and by experimentalists, even those who do their experimental work at large, national facilities. The individual researcher thus forms the backbone of research in condensed-matter physics, and the continued health of this area of physics is inextricably intertwined with the existence of adequate support for that backbone.

In what follows we make several recommendations directed toward accomplishing this. In doing so we recognize that a balance must be struck between the needs of researchers in their own laboratories and those of the external, user facilities that provide the equipment not available in individual laboratories. Both are real, both must be met, and there is a cost to each. The fact that research in condensed-matter physics is done primarily in small groups does not make it inexpensive. At the same time, the nature of the research equipment provided to users at the national facilities, and the staff needed to make it usable to visiting researchers, ensures its high cost to build and maintain. Because of large operating costs and the growing fraction of the condensed-matter community that uses them, user-oriented national facilities have the potential of drawing needed resources from the individual researchers in their own laboratories. *Both* needs must be met over the next decade. This will require the appropriation of enough new monies that the well-being of national facilities for users will not have a negative impact on those researchers who do not use them, and vice versa.

Support for Individual Researchers

MANPOWER

It is vital that there be a continuing stream of new scientists entering condensed-matter physics, to provide fresh ideas and to participate in the research that will keep the discipline lively and to train the students who will eventually replace them.

At present there are great opportunities for exciting work to be done in condensed-matter physics. But many physics departments find that there are more graduate students who are eager to pursue these opportunities than can be supported. This situation represents a lost opportunity to progress rapidly with the science as well as a lost opportunity to preserve and augment the nation's scientific and technological manpower in an area with important implications for the electronics industry and national security.

The national laboratories have performed an important educational role in the postdoctoral training of young researchers as well as the training of graduate students who have carried out their thesis research using laboratory facilities. It is important to maintain the ability of the laboratories to hire postdoctoral fellows. This is extremely difficult in the face of constricted budgets at the laboratories, and provision should be made for the continuation of the postdoctoral program.

Moreover, it is essential that there be opportunities for the most talented young investigators to obtain support for their research once they have left graduate school and have elected to pursue careers in a university setting. Otherwise science will decline because students will no longer enter the field.

It is also important that senior scientists who leave industrial or government laboratories to assume faculty positions in U.S. universities, as well as those who join our universities from abroad, have opportunities to obtain support for their research. Otherwise our universities will lose these sources of their enrichment.

It is equally essential that senior scholars doing first-class research at the frontiers of knowledge continue to have their work supported at adequate levels for the contributions that their work makes to knowledge itself.

However, as an illustration of recent trends, the Division of Materials Research of the National Science Foundation has found it necessary to decrease the number of grants (by about 20 percent over the past 5 years) to assure the viability of the best research programs through an increase in the average size of each grant awarded. Such a cutback in the size of the funded condensed-matter physics community should not be mistaken for quality control. The latter is provided by the turnover that occurs annually among grantees through the review of the proposals submitted to federal funding agencies. Diversity in the range of research activities shrinks with the reduction in the number of grants. The increasing competition for a decreasing number of grants gives rise to a tendency for investigators to be conservative in the submission of proposals by omitting speculative projects in favor of those that are almost guaranteed to be successful. This is not the way in which major advances in science are made.

Despite such efforts to increase the level of funding it is still the case that it is virtually impossible for an active research group to survive if its support comes from a single grant.

In order to restore support for those able scientists who have lost funding for this reason, and to provide resources to new scientists entering the field, it is important that there be an increase in the overall number of grants supporting research in condensed-matter physics. We believe that further cuts in the number of principal investigators currently supported would be inimical to our national interests. At the same time, it is clearly essential to be able to fund new young investigators, as well as to increase the funding of some currently supported young investigators as they develop into the major research directors of the next decade. Consequently,

We recommend that over the next 4 years sufficient new monies should be appropriated to provide for an annual increase of at least 3-4 percent in the *total* number of investigators in condensed-matter physics supported by the several federal funding agencies and for increases in grant sizes.

We stress that the purpose of this recommendation is to accelerate scientific progress, to exploit rapidly expanding opportunities, and to ensure continued availability of skilled manpower in a field whose health is essential for maintaining the vital flow of new technology.

INSTRUMENTATION

Crucial to an experimentalist's ability to perform experiments at the technical limits of a subject area is having equipment at the state-of-the-art level. Experience has shown that improvements in scientific instrumentation invariably lead to significant new discoveries. With the passage of time, and the steady improvement in the quality of experimental equipment commercially available, much of the instrumentation in university laboratories in the United States is no longer state of the art, at the same time that the cost of its replacement and upgrading by state-of-the-art equipment has increased substantially.

Recent studies show that the rate of increase in the cost of such equipment has averaged approximately 17 percent per year for the past several years. This is much higher than the rate of inflation of the consumer price index for the same period.

The changes in the directions of research in condensed-matter physics in the past few years have brought with them needs for qualitatively new kinds of experimental equipment that were not in existence when the decade began and that are costly. For example, an ultrahigh-vacuum system for research in surface physics may cost in excess of $200,000.

The study of the remarkable properties of artificially structured materials, such as semiconductor superlattices, requires having samples of such materials. Specialized equipment is necessary for their preparation, so that the cost of materials preparation in this field is high. A fully instrumented molecular-beam epitaxy unit can cost $1 million. Few universities have one. This means that university researchers are essentially shut out of some exceedingly important areas of research.

One of the significant changes that has occurred in the instrumentation needs of experimentalists during this period is their utilization today of powerful computational resources. Once an almost exclusive preserve of theorists, computers now control experiments and perform on-line data analysis. Qualitatively new capabilities have emerged in diverse applications such as on-line Fourier transformation for infrared spectroscopy or studies of chaotic processes, and image simulation for electron diffraction. Some experimentalists require only relatively small microcomputer systems, but these must be available for *each* experiment in the laboratory. Others require systems with sophisticated graphics and the ability to process large data files. Such systems are becoming available for about $50,000.

Accompanying the change in the kinds of instrumentation that are being, and will continue to be, used increasingly in condensed-matter research is the fact that in many cases more people are required to operate and maintain this equipment, increasing the cost of that research thereby.

In addition to making it possible for research at the highest levels to be conducted in these and other fields in our universities, there is another, educational, aspect to making such equipment available to university laboratories throughout the country. If the products of our graduate programs are to be able to step into positions in universities, industrial laboratories, or government laboratories where such equipment is being used and employ it productively, they must be trained in its use while graduate students.

This situation of aging instrumentation coupled with a shortage of renewal funds, and the problem it poses for the future of experimental science in the United States has been recognized in studies conducted recently by the National Science Foundation and the Association of American Universities, among others. In response to it, the Division of Materials Research (DMR) of the National Science Foundation (NSF) has established an ongoing instrumentation program that had $4 million budgeted for it in FY 1983 and $7.6 million in FY 1984, of which perhaps 40 percent goes for condensed-matter physics research. At the same time the Department of Defense (DOD) has begun a program to improve research instrumentation at universities, not just in condensed-matter physics, with $30 million in FY 1983 funds. It is intended that this initiative continue for 5 years, with $30 million budgeted in each fiscal year.

The response to these instrumentation programs was overwhelming. Approximately $27 million in requests was received by the DMR for $4 million of FY 1983 funds, while $750 million in requests was received by the DOD for $30 million of FY 1983 funds. Even allowing for the possibility that not all of the requests were of equally high priority, the magnitude of this request indicates the great need to upgrade or replace existing instrumentation in university laboratories. In the face of the magnitude of this need, the DOD and the NSF initiatives, while welcome indeed, are by themselves inadequate to meet it. It is, moreover, a need that must be met now: the longer a massive upgrading of scientific instrumentation is delayed, the larger becomes the degree of obsolescence of equipment in U.S. laboratories and the greater the overall cost of its replacement. Additional funds for this purpose are urgently needed. Consequently,

We recommend that sufficient new monies be appropriated to enable federal agencies supporting condensed-matter research to dedicate a certain continuing portion of their support dollars to instrumentation programs. At present levels of support we recommend that this portion be of the order of 20-25 percent.

We believe that over the next 4 to 5 years U.S. laboratories will be significantly revitalized thereby. It is important, however, that support for instrumentation be provided on a continuing basis in the future in order that U.S. laboratories remain in the forefront of research.

COMPUTATION

Almost all areas of this report confirm that computers play a vital and increasing role in condensed-matter physics. Therefore this trend must be recognized by funding agencies and appropriate provision made for the purchase and maintenance of the special computer systems required for research progress.

Computers have had an enormous impact on both the speed and the kinds of condensed-matter calculations that can now be done. They have become theoretical laboratories for dynamical simulations of condensed systems and for their statistical analysis by classical or quantum Monte Carlo methods. Specifically configured computers are now being designed and built to address particular types of theoretical problems. This increased use of computers by theorists is making present-day research in condensed-matter physics more rewarding but also more expensive. A supercomputer, such as the CRAY-XMP, costs around $10 million, and at present only three universities in the United States have one. A computer of the type of a VAX 11-780 costs in the vicinity of $200,000. Even if research groups do not own their own computer, the cost of purchasing computer time can be a significant portion of the cost of present-day research. The future health of the field requires that this clearly identified need for computers be accommodated.

As computing needs are diverse—requiring both increased capacity and capability—they cannot all be filled by the medium-sized mainframe computers found on most research campuses. There is a specific need in theoretical work for advanced computing capabilities. Some of these can often be met by adding a fast processor to a conventional computer at a cost of about $400,000. Almost an order of magnitude or more additional computing power can be provided by modern supercomputers and, in some fraction of cases, the provision of time on such machines—if not the machines themselves—for condensed-matter research is essential. The future funding patterns must accommodate the growing demands for computer use. To this end,

We recommend that sufficient new monies be appropriated to allow the several federal agencies supporting condensed-matter research to identify a continuing fraction of the total budget to be devoted to the special computing needs of condensed-matter research. The assignment to computing of 10 percent or more of the total present budget would appear well justified. The funds so assigned could be used for the purchase of computer time or for the purchase and maintenance of dedicated equipment.

In view of rapid changes in computer technology and patterns of use in the physics community, this fraction should be reconsidered after the next few years of experience. The scientific community and the federal funding agencies should work together to promote more effective use of major computer resources through networking, standardization, and the establishment of user assistance groups.

FUNDING

The chances of realizing the research opportunities that the coming decade offers will be significantly enhanced by an increase in the number of individuals carrying out this research, an increase in the level of support that they receive, and the provision of the increasingly more sophisticated, and the increasingly more costly, equipment that they will need in their work.

The costs of conducting a modern research program include the maintenance of equipment, the operating or running costs, and the salaries and benefits for personnel. These costs will increase more rapidly than inflation because of the increased sophistication of the required equipment and the expected increases in tuition for graduate students.

The national investment required for the adequate support of basic research in condensed-matter physics by individual researchers, however, is not great, even though the return on the investment is large. There is heartening evidence in the current federal budget, through its approximately 20 percent increase in support for basic science over the level for last year, that this is recognized. However, more still needs to be done to capitalize on the opportunities that exist.

We estimate that implementing the preceding recommendations will require an increase in funding for research in condensed-matter physics at a steady annual expansion rate of approximately 20 percent in constant dollars for an additional 3 years. We strongly recommend that this increase take place.

The special claim of condensed-matter physics for research support from federal and industrial sources lies in its record of converting deep science into benign and sophisticated industrial technology on a time scale that is often no more than 5 to 10 years. This process is still vigorously under way with such notable new scientific discoveries as the quantized Hall effect, valence fluctuations, heavy electron-mass metals, electron localization due to disorder, artificially structured materials, conducting polymers, chaotic phenomena in solids and liquids, and solitary wave phenomena in solids. If the resources become available to carry out the research necessary to exploit these new discoveries, the impact on industrial technology will be even greater than what has gone before.

Support for National Facilities

Some of the national facilities are comparatively new; others have been in existence for many years. Because of their importance for the nation's scientific effort, the facilities that continue to maintain a high level of scientific excellence should be adequately supported. Planning for new facilities to meet the needs of new areas of condensed-matter physics that are now developing must begin in the near future.

The needs of the neutron and synchrotron facilities have been subjected to detailed scrutiny recently by several panels sponsored by the NSF and the Department of Energy (DOE). The most recent of these studies* was prepared while this report was being written. We will have occasion to refer to it in what follows.

NEUTRON FACILITIES

The existing high-flux reactors, the cornerstones of the U.S. neutron-scattering program, are underfunded and understaffed. Relative to their Western European counterparts they are falling seriously behind in instrumentation. Therefore,

We recommend that a concerted and coordinated effort should be undertaken to expand the effectiveness of our high-performance reactors by adding new, diversified instruments along with personnel necessary to design, build, and utilize them in the user mode. We estimate that at least ten new instruments are needed, requiring an increase in annual operating costs of $2 million to $3 million for manpower needed for their design and use. About $20 million to $30 million is required for building such instruments, to be spent over 5 to 7 years. Instrumentation plans beyond the level projected above may be warranted but should be justified by demonstrated user needs.

Note that this estimate does not attempt to address the somewhat different needs of the chemistry and biology communities. A 1984 Panel on Neutron Scattering, considering the total scientific community, estimated a need for ~30 new instruments.†

Spallation sources provide new opportunities to expand the power of the neutron as a probe of condensed matter. The United States currently has two pulsed spallation sources. The Los Alamos Neutron Scattering Center (LANSCE) facility at the Los Alamos National Laboratory is compromised currently by the pulse structure of the LAMPF proton beam that supplies it. This situation will be corrected by the addition of a proton storage ring (PSR) scheduled for completion in 1986. It is also restricted by the small experimental hall. The Intense Pulsed Neutron Source (IPNS) at the Argonne National Laboratory, with an active outside-user community, an experienced staff, and an adequate experimental hall, is the highest-performance source in operation at present.

---

* *Major Facilities for Materials Research and Related Disciplines* (National Academy Press, Washington, D.C., 1984). This will be referred to below as the report of the Seitz-Eastman committee.

† *Current Status of Neutron-Scattering Research and Facilities in the United States* (National Academy Press, Washington, D.C., 1984).

We therefore recommend that funds be appropriated to enlarge the LANSCE instrument hall (a $15 million construction project has been proposed) and operation of IPNS be continued until the latter's ongoing activities can be accommodated by a more powerful and cost-effective LANSCE, provided this can be budgeted without jeopardizing the necessary rejuvenation of the high-performance reactors.

Very recently Argonne has proposed the upgrade of IPNS by the replacement of the existing accelerator with one of new design (fixed field, alternating gradient) with a sevenfold increase in proton current. If this design is shown to be practical and cost-effective relative to LANSCE, it will be necessary to reconsider our spallation-source priorities in the light of the existing investments.

There are no comprehensive plans at present concerning the status of our neutron capabilities for the 1990s. Given the uncertainties in the lifetimes of existing facilities and the time necessary for the design and construction of new facilities, it seems advisable for the neutron-scattering community to initiate discussions immediately leading to such a plan. The feasibility and desirability of both steady-state and pulsed sources should be studied. The possibility of establishing such a facility through international cooperation should also be fully explored.

We therefore recommend that supplemental funds be made available to interested qualified institutions to investigate various options for an advanced neutron source. These studies should be done in parallel and in consultation with a panel of outside users charged with devising a plan that will ensure that our neutron-scattering needs will be met in the 1990s and beyond.

SYNCHROTRON RADIATION SOURCES RECOMMENDATIONS

Synchrotron radiation has had a broad impact on studies of both the structural and electronic properties of condensed matter. This is due to its unique high brightness, wide tunability, high polarization, and narrow angular divergence (and, in some instances, time structure). These properties are similar to those of laser sources, but the wavelength range of synchrotron radiation extends from that of the shortest known laser wavelength throughout the ultraviolet, soft-x-ray, and hard-x-ray regions.

It is recommended that the current new generation of synchrotron facilities be completed as soon as possible since their high brightness will serve the short-term needs of the next 3 to 5 years.

The main scientific emphasis of these short-term objectives should be in the following areas: (i) Current beam-line instrumentation should be refined in order to achieve higher resolution of photon monochromators in the conventional VUV (0-100 eV) and x-ray (4-15 keV) ranges. This will allow new types of studies to be made of electronic and structural phenomena in conventional solids as well as in low-dimensional systems such as surfaces, polymers, and liquid crystals. (ii) Novel new instrumentation should be developed for soft x rays in the 100-4000 eV range that uses combinations of conventional diffraction-grating technology with new synthetic materials such as multilayer mirrors and other x-ray optical elements. This would allow high-resolution studies of the shallow core-level spectra of all elements. In addition both extended-x-ray absorption fine-structure (EXAFS) and high-resolution near-edge studies could be performed using K or L edges of elements with an atomic number smaller than that of xenon. In order to exploit the potential of insertion devices in the x-ray region it is important that the design allow first harmonic undulator radiation at energies up to ~20 keV.

A commitment should begin immediately toward the next generation of high-brightness synchrotron facilities using insertion devices. This should be a two-step approach.

New undulator and wiggler devices should be constructed on existing storage rings so that insertion-device technology will move ahead rapidly and be ready for possible new rings. New optical devices should be developed to match insertion device sources; this should be done in parallel with the development of new sources, since higher resolution and wider tunability cannot be achieved simply by attachment of existing beam lines to new sources.

As a second priority, planning should begin immediately leading to proposals for a next-generation, possibly all-insertion device machine. Ideally, this machine should be completed in the early 1990s, since projected user demand will saturate then-existing facilities by that time. The design parameters, such as electron energy and physical size, should be determined by scientific considerations, but the three areas of spectroscopy, scattering, and microscopy should be accommodated. The 6-GeV machine recommended by the Seitz-Eastman committee appears to meet these needs. The overall costs of such a next-generation synchrotron source are in the range of $160 million, and construction could take place over a period of 6-7 years. Firm decisions on when to build such a machine should be made on the basis of new scientific opportunities, user demand, and ongoing experience with the undulator and wiggler facilities discussed above.

HIGH-MAGNETIC-FIELD FACILITIES RECOMMENDATIONS

Laboratories for the production of high magnetic fields (>15 T, where 1 T $\equiv 10^4$ Oe) and their utilization in condensed-matter research exist in France, Holland, Belgium, Japan, Poland, the Soviet Union, and the United States. The National Magnet Laboratory at the Massachusetts Institute of Technology is the only major user facility for high-field research in the United States. A wide variety of steady-field magnets exist there and are categorized by their peak fields, bore sizes, and homogeneity. The largest field currently available there is 29 T, in a 3.3-cm-bore hybrid configuration.

Magnetic fields above 30 T are economically feasible only in pulsed operation. Nondestructive, repetitive pulsed fields in the range 40 T $\leq$ $H \leq 60$ T are now available in Holland, Japan, and the Soviet Union. A 75-T configuration will soon be operating in Osaka, Japan. A high-magnetic-field facility has just been completed at the Institute for Solid State Physics (ISSP) in Tokyo, Japan, at a cost of about $10 million. It can produce a variety of nondestructive pulsed fields (≤ 50 T); it can produce fields of 50-100 T by plasma compression that may be nondestructive; and it can produce a 100-500 T implosion-generated field that is totally destructive of the sample. The ISSP group has been generating fields of 100 T for several years, which have been used in studies of cyclotron resonance and various other phenomena in semiconductors. No comparable facilities are available in the United States, although much of the seminal technology was developed in this country.

The availability of high magnetic fields has yielded such experimental results as the discovery of the fractionally quantized Hall effect. More generally, high-field magnets expand the phase diagram of a solid by adding a new variable, the magnetic field, to the usual variables, pressure and temperature, thereby increasing our knowledge of properties of solids under extreme conditions. For these reasons, and the paucity of high-field magnets in the United States,

We recommend that new money should be made available to enable greater emphasis to be placed on the generation of pulsed high magnetic fields at the National Magnet Laboratory and/or at a new site elsewhere in the United States. The cost of duplicating the high-magnetic-field facility in Osaka is estimated to be $1 million to $2 million.

ELECTRON-MICROSCOPE FACILITIES RECOMMENDATIONS

The country's electron-microscope facilities provide a reservoir of talent and expertise necessary to generate the innovative instrumentation crucial for promoting the growth of the power and subtlety of electron-microscopic investigations in the coming decade. There appear to be four major areas in which advanced instrumental initiatives could have a major impact on the development of the field during this period: (1) development of ultrahigh-vacuum sample environments for surface studies; (2) development of efficient instrumental accessories for microanalytical techniques such as electron energy loss spectroscopy; (3) development of low-temperature specimen stages and specimen preparation techniques necessary for systematically attacking questions about the structures of large biological molecules and of many others that are of interest to condensed-matter physics; and (4) development of computerized data collection and analysis. It is estimated that the cost of the major capital equipment required for implementing these instrumental initiatives would average $1 million for each, spread out over a period of 2 years, for a total of $4 million. The increase in the operating budgets of the institutions participating in these initiatives is estimated to be $4 million, to be achieved over a period of 3-4 years. Our recommendation in this area is as follows:

Advanced instrumentation initiatives in the four areas of electron microscopy cited above should be established in response to competitive proposals from interested institutions. If necessary, the federal funding agencies should stimulate the submission of such proposals.

GENERAL RECOMMENDATIONS CONCERNING NATIONAL FACILITIES

There are two broad categories of users of national facilities. Committed users are those whose research programs are built nearly exclusively around the use of these facilities and include the scientific staff of the facilities. By contrast, occasional users have research programs based on other techniques, usually at their home laboratories, but whose research is increased in scope by the power of these other specialized techniques. The long-term vitality and future growth of national facilities depend crucially on a broad base of these occasional users who have neither the time nor the financial resources to become expert in these techniques but who furnish nonetheless a wealth of novel materials and ideas for experiments. In order to aid the integration of these occasional users into the activities of the facilities,

We recommend that special funding be set aside for the purpose of accommodating occasional users at the national facilities. This money would help finance travel and living expenses, particularly for university users, and provide an increase in the in-house support staff. This program should be formulated by the individual facilities in consultation with university and industrial collaborators and funded on the basis of separate proposals from these facilities. We estimate that a significant trial program would require $4 million to $6 million per year over a 3-4 year period.

Recently established national facilities (e.g., those dedicated to research employing synchrotron radiation or high-resolution electron microscopes) have been developed as user facilities or as DOE Centers for Collaborative Research. The independent peer review of the experiments approved to be done improves the quality and nature of the research at these facilities. At the same time, the ability to respond to rapidly emerging scientific opportunities and the timely development of new experimental techniques requires that a certain fraction (perhaps 30 percent) of the available time be allocated at the discretion of the in-house staff. Therefore,

We recommend that in the future it is desirable that national facilities should operate in the user mode in which the majority of experimental time is allocated by independent peer review.

There are at least two modes in which this peer review may operate: review of experiment-by-experiment proposals by occasional users and peer review of proposals for participating research teams (PRTs) that undertake to construct, maintain, and carry out research programs using instruments on a shared basis with non-PRT members.

Finally, it is our strongly held view that the needs of the individual researcher, which have been outlined above in the section on Support for Individual Researchers, are so great at this time that the highest priority for the use of new monies for the support of condensed-matter physics is in meeting those needs and for the upgrading of the existing national facilities that is necessary for the achievement of their full potential. When this has been accomplished, the construction of the new national facilities should begin.

University-Industry-Government Relations

One of the primary strengths of condensed-matter physics is that forefront research of the highest quality is carried out at industrial laboratories as well as at universities and government laboratories. This is due to the fact that condensed-matter physics is closest to applications in technology of all the subfields of physics. It argues for a strong coupling between universities and national laboratories, where most of the basic research in condensed-matter physics is done, and industry, where the results of that research, as well as of the research done in-house, is transferred into technology. Industry also benefits greatly from the pool of condensed-matter physicists produced each year by this country's universities and from those trained in postdoctoral programs at national laboratories. For their part, universities have received support from industry in the form of grants of equipment, funding for research projects, and support for graduate students. However, if the strongest possible coupling between universities, government laboratories, and industry is to be achieved, the support of university and laboratory research by industry should go well beyond the mere provision of funds and equipment: research *cooperation* is also required. At the same time, continuing efforts should be made to increase the research cooperation between the national laboratories and university scientists, since special facilities exist at the national laboratories that are not available elsewhere. The realization of such cooperation will require the coordinated efforts of universities and industry, and of the federal government as well. The following recommendations outline our views of the roles of each of these partners in this process.

1. What government should do:

Establish policies, including tax incentives, to stimulate fundamental research in industry.

Provide support for students engaged in cooperative university-industry research.

Encourage and facilitate the flow of scientists between federal laboratories and universities for cooperative research programs.

Maximize access by outside users to the special facilities available only at the federal laboratories.

2. What industry should do:

Increase the amount of in-house research even beyond the levels directly supported by the policies suggested in point 1 above, i.e., through the use of corporate funds.

Establish and fund programs that enable industrial scientists to take sabbatical leaves in universities and at national laboratories.

Receive university faculty and laboratory researchers in industrial laboratories for sabbatical leaves and summers.

Provide direct support of faculty and departmental research grants (e.g., the IBM programs).

Provide direct support of graduate and postdoctoral fellowships (e.g., the IBM fellowship program).

Formulate cooperative research projects with graduate students (e.g., the MIT-AT&T Bell Laboratories program).

Provide instrumentation for special facilities at national laboratories.

3. What universities should do:

Implement cooperative research and support programs with industry, as MIT has done in materials processing.

Adopt a limited form of the "Japanese model" in which applied physics research in high-technology areas, such as semiconducting lasers, photonics, and electronics is supported by industrial firms directly involved in the manufacture of materials, devices, components, and systems employing these technologies.

Cooperate in the graduate training of industrial employees engaged in applied research.

Arrange for sabbatical leave for federal laboratory researchers in university departments. This support can take the form of direct research contracts; the gift or loan of equipment, devices, and components; and the support of graduate students.

II

A Decade of Condensed-Matter Physics

In this part we summarize some of the advances in condensed-matter physics in the past decade, including significant and interesting milestones that will ensure the continuing liveliness of the discipline. These summaries are accompanied by descriptions of areas that provide particularly exciting research opportunities in the next decade.

The division of condensed-matter physics into the subareas represented by the following chapters was made because it corresponds to the communities of scientists who relate most directly with each other, for example through the organization of workshops and international conferences. It consists roughly of two broad categories devoted, respectively, to solids and to liquids. The distinction between these two phases can be made on the basis of their symmetry properties, i.e., on whether they are left unchanged under uniform displacements (translations) and rotations. Chapters 1-7 are devoted to solids, and the discussion of liquids that follows is prefaced by a chapter on quantum fluids, which span both the solid and the liquid state. The importance of new experimental techniques, laser spectroscopy, new materials, and the national facilities to the advances that have been made, and can yet be made, in the subareas of condensed-matter physics is delineated in the appendixes. Finally, the emphasis in what follows is on the discoveries and opportunities: no attempt was made to identify the individuals responsible for the discoveries or their institutions.

1

Electronic Structure and Properties of Matter

INTRODUCTION

Electronic structure and the properties of matter is a vast topic that is at the heart of all condensed-matter physics. It might be described as the electronic quantum many-body problem and is concerned with the ways in which the effects of the Pauli exclusion principle and the Coulomb interactions between electrons conspire to produce the remarkable varieties of matter. During the last decade, concerted efforts were made to determine the most efficient means of incorporating the effects of exchange and correlation into the basic description of solids and liquids, with the result that significant advances have occurred in our understanding of the electronic structure of large systems with perfect order, with various types of defects, and with disorder, including both liquid and amorphous states.

This period has also seen great strides in our understanding of the surfaces of condensed matter and the properties of interfaces. In addition, our attention has turned to systems of unusual chemical character, quasi-one- or two-dimensional solids, for example, with physical properties often remarkably different from those of the higher symmetry three-dimensional systems that have so influenced the development of condensed-matter physics. These low-dimensional materials demonstrate the effects of electron-electron interactions in the most dramatic way. The resulting electronic order can manifest itself in magnetically ordered states, in superconducting states, or in charge-density waves associated with unusual spatial structures, in the fractionally quantized Hall effect, and in many other new phenomena. These systems clearly demonstrate that the synthesis, characterization, and analysis of new materials may be expected to continue to lead to discoveries of fundamental significance.

ADVANCES IN ELECTRONIC STRUCTURE DETERMINATIONS

For simple systems of relatively high symmetry it is now possible, with little more than the atomic number of an element as primary input, to account for their major ground-state properties, such as the lattice structure, lattice constant, bulk modulus, and density. No information peculiar to the condensed state is used at all.

The basis of this advance is a progressive acceptance of the density-functional method for treating exchange and correlation in the electronic ground state. This method utilizes the existence of a certain functional of the electron density and its gradients, and the Coulomb interactions, and kinetic quantum energies as the basis for constructing the free energy of an electron system. Although the functional is determined only from properties of the uniform, interacting electron gas, it is widely used in cases for which the electron density is grossly inhomogeneous, such as in crystals (Chapter 2) and on surfaces (Chapters 5 and 7). With the use of appropriate spin-polarized functionals it is also possible to study magnetically ordered states (Chapter 4). Together with the development of methods for calculating electronic states—the first-principles pseudopotential, linearized muffin-tin orbital, and linearized augmented-plane-wave methods—the density functional method has been used to study ground-state properties of a wide range of disparate systems. Band-theoretic methods of this kind are impressively predictive. In the near future they are expected to be applied to more complex real-space structures, to ionic and partially ionic systems where difficulties in its application still remain, and to the technologically important problems of interfaces.

More challenging yet is the physics of the excited states of such systems. Their properties are directly probed by powerful techniques such as angle-resolved photoemission or radiation (ultraviolet and x-ray) from synchrotron sources. With ordinary photoemission, the so-called *angle-integrated* or energy-resolved photoemission, radiation impinging on the surface of a sample excites electrons in its interior to higher energy bands. Some of the excited electrons then move to the surface and tunnel through it to the exterior, where they are detected.

If selected *only* according to their final energy, the electrons give information, under certain conditions, about the joint density of states of the energy bands from which they were originally excited. Under other conditions, the electrons can also give information about the surface itself, and even about surface overlayer atoms deliberately adsorbed onto it.

With *angle-resolved* techniques, the emergent electrons are selected not only according to energy but also as to direction or wave vector. With this new information it is possible to map out the energy-band structure itself and not merely the joint density of states. The experimental results can then be directly compared with calculated energy bands, providing information on the electronic structure of a given material. It is also possible to map out rather detailed features of the bulk-energy-band structures, which previously were not available from more traditional probes (e.g., optical excitation and interband absorption). This information, especially for higher bands, is nicely complemented by data obtained from brehmsstrahlung isochromat spectroscopy (often referred to as inverse photoemission).

The band-theory density-functional methods can in principle be adapted and extended to excited states as well. This is an area of great current activity, where there is as yet no solution to such basic problems as obtaining the fundamental band gaps in crystalline semiconductors accurately.

In summary, it is clear that some aspects of electron structure in ordered systems are quite well understood, to the point where application of theory to materials exhibiting unusual properties (the high-temperature, superconducting A-15 compounds, for example) leads to further suggestions for exploiting the particular properties of interest. Other aspects, particularly involving the many-body problems of electronic excited states, are not understood and are currently the subject of much interest and controversy.

MANY-ELECTRON EFFECTS

The central task of those interested in electronic properties is to understand a problem involving an immense number ($\sim 10^{23}$) of strongly interacting electrons. The historical approach to it was to begin by treating each electron as actually *independent* of its peers. Electron-electron interactions were not completely ignored but were treated in some average sense.

It has been well known for many years that a large number of problems in condensed-matter physics require going considerably beyond this *independent* electron approximation. One such problem that has received much attention during the past 10 years concerns the way in which an electron from the interacting-electron system can make a transition to a prepared, "deep," atomic state of one of the atoms. In doing so, the electron emits an x ray that can be detected. As well as giving information on the width of the band from which the electron fell, the x-ray intensity probes the effects of electron-electron repulsion in the dense, interacting electron system. In particular, the edge structure, reflecting the onset of the radiation, is dramatically affected (the so-called x-ray edge problem).

This problem has been treated using the extremely powerful renormalization group techniques (see Chapter 3) in which short-wavelength degrees of freedom are systematically replaced by averages over successively larger length scales. This method, remarkable in its accuracy and generality, involves extremely creative use of the computer.

Another recent development in the study of many-electron effects is the use of statistical Monte Carlo methods to find accurate numerical solutions of the full many-body problem. These methods are in essence computer simulations of systems containing finite, but large, numbers of interacting particles. (The name comes from the use of random numbers that could be generated on a small scale by a roulette wheel but that are produced by complicated mathematical algorithms for research studies.) The applications of these methods have included obtaining the thermodynamic properties of condensed matter, such as the equation of state of solid and liquid phases. Recently these methods have been extended to quantum-mechanical problems and applied to liquid and solid ^{4}He, considered as bosons interacting via realistic potentials. They have yielded excellent agreement with experiment for the equation of state and for the probability distribution of the distances between atoms in the liquid. Related methods have been applied to other systems, such as electrons in one dimension interacting with each other and with a lattice. This has made possible a rigorous study of phase transitions that takes into account the full quantum fluctuations present in one dimension and has revealed important differences from mean-field descriptions of the transitions. These developments have also led to Monte Carlo techniques for the determination of properties of fermion systems, which are difficult to calculate because of the antisymmetry of the wave function.

QUANTIZED HALL EFFECT

One of the most surprising recent developments in condensed-matter physics has been the discovery of a set of phenomena collectively known as the quantized Hall effect. These phenomena are associated with two-dimensional electron systems, in a strong magnetic field and at low temperatures. In practice, the electron systems studied are semiconductor inversion layers, such as occur in a metal-oxide-semiconductor (silicon, for example) field-effect transistor (MOS-FET) or at a GaAs-GaAlAs heterojunction.

When a current-carrying wire is placed in a steady magnetic field (Figure 1.1), a voltage V_H is developed across the wire in direct proportion to the current density. This well-known phenomenon is the Hall effect, and the voltage V_H across the wire is known as the Hall voltage. Classically, the Hall resistance R_H, defined as the ratio of V_H to the current I, is expected to vary linearly with the applied magnetic field and inversely with the carrier concentration in the sample. For two-dimensional electron systems at very low temperatures, however, the Hall resistance was found to exhibit a series of plateaus, with varying carrier concentration or magnetic field, and the value of Hall conductance ($1/R_H$) on these plateaus was found to be quantized in precise integer multiples of the fundamental unit e^2/h, where e is the electron charge and h is Planck's constant. (The resistance h/e^2 has the

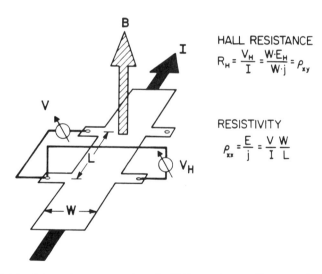

HALL RESISTANCE

$$R_H = \frac{V_H}{I} = \frac{W \cdot E_H}{W \cdot j} = \rho_{xy}$$

RESISTIVITY

$$\rho_{xx} = \frac{E}{j} = \frac{V}{I}\frac{W}{L}$$

FIGURE 1.1   Schematic representation of a Hall experiment. The magnetic field B is perpendicular to the plane of the specimen and to the current I. The Hall resistance R_H and the resistivity ρ_{xx} are determined through the equations shown in the figure. [Courtesy of H. L. Stormer, AT&T Bell Laboratories, Murray Hill, New Jersey; adapted from K. von Klitzing, G. Dorda, and M. Pepper, *Phys. Rev. Lett.* 45, 494 (1980).]

value of 25,812.8 ohms.) Moreover, when a Hall plateau occurs, the voltage drop *parallel* to the current is essentially found to vanish, so that the current appears to flow through the sample with no observable dissipation (see Figure 1.2). The surprising precision with which e^2/h can be measured in this way may lead to a new, practical, secondary resistance standard and to an improvement in the determination of the fundamental constants. The precision of the effect (now established to about 2 parts in 10^8) is observed in spite of considerable variations in sample properties, and this has led theorists to propose fundamental explanations of the effect. The discovery of the quantized Hall effect was honored by the award of the Nobel Prize in 1985.

In 1982, Hall conductance plateaus at certain simple *fractions* of the quantum e^2/h were discovered in GaAs heterojunctions of exceptionally high mobility, in magnetic fields so high that the first magnetic quantum level is fractionally occupied. These results are perhaps even more remarkable and surprising than the original observations of integral, quantized Hall plateaus. Although the integral steps had been explained in terms of the quantum states of individual electrons, ex-

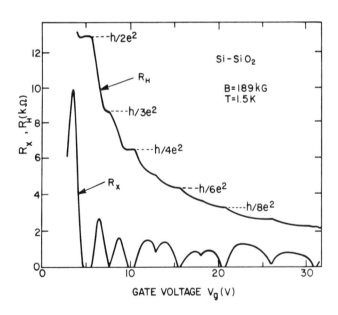

GATE VOLTAGE V_g (V)

FIGURE 1.2 The quantized Hall effect in a Si MOS-FET in which the electron density is varied by a gate voltage V_g. Instead of being a smooth curve, the Hall resistance R_H develops plateaus having values h/ie^2, where i is an integer and the resistance R_x of the specimen drops to low values.

planation of the new fractionally quantized Hall effect has required the hypothesis of a radically new type of quantum liquid state for the collective motion of electrons in a magnetic field. Many aspects of these systems of electrons in strong magnetic fields are still poorly understood, and this will certainly be an active area of research in the 1980s.

ELECTRON-HOLE DROPLETS

The conductivity of semiconductors arises from thermal excitation of electrons, or holes, from the bands or bound impurity levels that are normally filled at low or near-zero temperatures. However, by the use of intense laser radiation, immense numbers of electrons can be excited from lower-lying states, leaving behind the absent electrons (i.e., the hole states). The resulting nonequilibrium populations of electrons and holes can be formed quickly. Although they are initially dispersed, the electrons and holes rapidly partially equilibrate into a new state that consists of electron-hole droplets (Figure 1.3). The experimental signature of their existence is that, when the electrons do eventually return to their lowest-energy states, the distribution of radiation emitted is characteristic of the condensed Fermi seas, representing the arrangement of excited electrons and holes. During the last few years, the formation of electronlike droplets and their essential characteristics have become far better understood. It is known that band structure, many-electron effects, and specifically correlation effects also enter in an essential way so that this phenomenon has led to a substantially improved understanding of interacting electron systems.

ELECTRONICALLY ORDERED STATES

Historically, the paradigm of an electronically ordered system is a substance exhibiting one of the forms of magnetic order. The later discovery of superconductivity is another dramatic form of electronic order: these topics are discussed in Chapters 4 and 8, respectively. Both are now being viewed more broadly, especially in terms of their bearing on other manifestations of electronic order, some of which have been discovered quite recently. Important to the discovery of these new forms of order has been the fabrication, characterization, and analysis of new materials, especially those that exhibit different states of order as temperature is altered (see Appendix C).

In the context of ordered electron states and the connection with atomic arrangement, the dimensionality of the system, once again, plays a crucial role in the physics. Thus, we now find quasi-one-

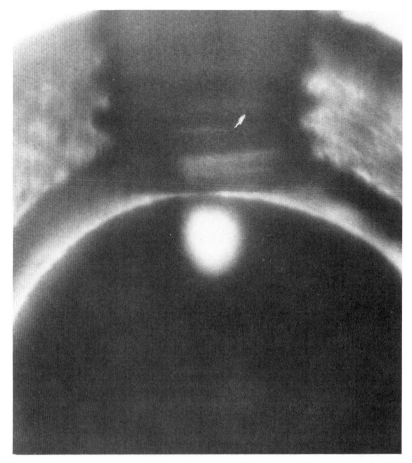

FIGURE 1.3   Photograph of the 1.75-μm recombination luminescence emanating from a strain-confined drop of electron-hole liquid. The 4-mm-diameter disk of ultrapure Ge is pressed along (110) at the top with a nylon screw, creating a stress maximum inside the crystal. The liquid is a degenerate sea of electrons and holes with a density of $\sim 10^{17}$ cm^{-3}. (Courtesy of Carson Jeffries, University of California, Berkeley.)

dimensional materials whose constituent atoms are not drawn from the metals but that nevertheless do behave, quite remarkably, as metals. A striking example is polyacetylene, a quasi-one-dimensional system that is formed by catalytic polymerization of acetylene gas. It is apparently the first organic polymer to be made to conduct and joins materials such as tetrathiafulvalene-tetracyanoquinodimethane (TTF-TCNQ) as examples of linear metals (see Chapter 10).

A particularly interesting category of materials is the class of layered compounds, examples of which are TaS_2, $TaSe_2$, and $NbSe_2$. The name comes from their tendency to group atoms in planar, sandwich arrays. Most compounds with relative chemical simplicity crystallize into regular and often relatively simple structures. In some, however (certain transition-metal chalcogenides), and especially at low temperatures, structural instabilities are observed to result from the interplay of the interactions among electrons and the interactions among the electrons and the ions of the material. The ensuing states may then display modulations in charge density, modulations in spin density, or even modulations in the ion density. What is especially fascinating is that the resulting systems are no longer truly periodic. In the two-dimensional or layered compounds, experiments have revealed the curious feature that the period of these modulations can actually be incommensurate with the fundamental repeat distance of the original underlying lattice (see Figure 1.4).

The energy balance in these systems is such that relatively modest changes in temperature can cause changes in the states of order. This is also true in another interesting class of materials formed by interposing (or intercalating) atoms of certain substances between the planes of graphite crystals. An example is graphite intercalated with cesium. It is possible to stage such materials, i.e., to interpose a fixed number of layers of the host system between consecutive intercalate planes (see Figure 1.5), and the resulting systems exhibit a variety of interesting phase transitions.

This competition among possible orderings also exists in linear systems as well: thus, for example, $(SN)_x$ has been observed to be superconducting.

DISORDERED SYSTEMS

The study of electronic states in systems that do not have long-range order has become of increasing importance in the understanding of condensed matter. The most striking phenomenon in disordered systems is the localization of the true quantum-mechanical eigenstates. Localization by disorder alone, the *Anderson transition*, occurs when fluctuations in the potential associated with the disorder exceed a particular value. Close to, but below, this value, it is believed that there exists a mobility edge, at which energy the states change character from localized to delocalized.

The various manifestations of disorder are being probed experimentally in a variety of systems by photoemission, photoluminescence,

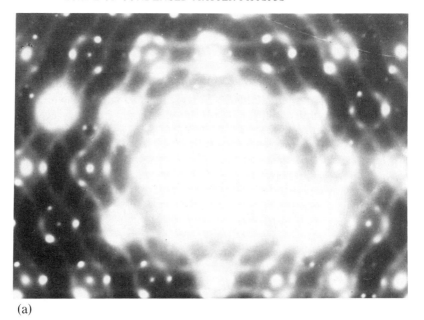

(a)

(b)

(c)

FIGURE 1.4   X-ray diffraction patterns of charge-density wave-bearing layered compounds. (a) 1T-$TaSe_2$ just above the commensurate-incommensurate phase transition at 473 K, in the incommensurate state. (b) $\sqrt{13}a_0$ superlattice of 1T-$TaSe_2$ in the commensurate charge-density wave state at room temperature. (c) 4Hb-$TaSe_2$ at room temperature in the commensurate state, producing the $13a_0$ superlattice. (Courtesy of F. J. DiSalvo, AT&T Bell Laboratories, Murray Hill, New Jersey.)

optical response, soft-x-ray emission, phonon echo, extended x-ray absorption fine structure, and many other techniques. The metal-insulator transition (the precipitous drop in conductivity itself) has been unambiguously reported, at low temperatures, in doped semiconductor systems. Furthermore, the behavior of the conductivity near the transition is related to critical phenomena in phase transitions. There is a growing realization of the importance of long-range Coulomb inter-actions in disordered as well as ordered systems. They can contribute to a transition from an insulating to metallic state or vice versa.

In contrast to Anderson localization, in the picture associated with the *Mott transition*, the view is that the electrons in the system can only be *cooperatively mobile* to the extent that the Coulomb interac-tions, included through screening, act to reduce the possibility of *cooperative recombination* with the charge centers from which they originate. Accordingly, as the mean-charge-center separation in-creases, the electronic bandwidths do not shrink continuously to zero but instead vanish suddenly at a certain critical density. If the system contains more than one electron per atom, then two or more *Hubbard* bands can be made to overlap as the lattice constant decreases. The effects may be particularly subtle and interesting in disordered sys-tems, and the interplay between correlation effects and the effects of disorder itself is an emerging area of research.

One aspect of the metal-insulator transition where there has been much theoretical and experimental progress in the last few years has been the study of localization phenomena in one- and two-dimensional systems—in particular the exploration of a class of phenomena known as weak localization. For example, it has been proposed that for a

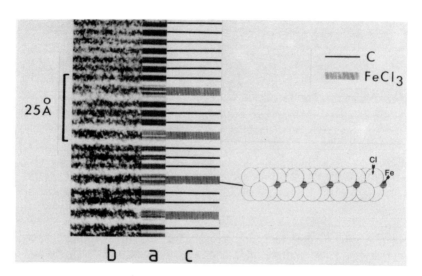

FIGURE 1.5   High-resolution electron micrograph showing the existence of mixed staging in ferric chloride graphite intercalate. Clearly shown are regions with two and three layers of graphite between the layers of $FeCl_3$. (Courtesy of John M. Thomas and coworkers, University of Cambridge.)

sufficiently thin wire at very low temperatures the resistance of the wire will no longer depend linearly on the length of the system. (In practice, to observe this departure from common behavior it is necessary to fabricate wires with diameters $\lesssim 100$ Å.) Similarly, the electrical resistance of a two-dimensional system is expected to increase logarithmically as the temperature is lowered. The theoretical methods used in making these predictions have included renormalization group techniques and scaling ideas similar to those used in the theory of critical phenomena (Chapter 3).

Interpretation of weak localization experiments is complicated by subtle effects of electron-electron interactions and of the presence of impurities with local magnetic moments or with strong spin-orbit coupling; moreover, the effects are small in practice and require precise low-temperature measurements. Nevertheless, the effects have been observed, and the dependence on temperature, on magnetic field, and on other parameters has been found to follow theoretical predictions rather closely in many cases. Flux quantization, which is expected in superconducting systems, has also been predicted and observed in normal metallic systems (Figure 1.6). These observations provide convincing evidence that the physical basis underlying the theory of localization is correct.

There are also important questions connected with the general nature of electron states in systems with weaker disorder. These can differ considerably from electron states in their crystalline counterparts. A striking example is Si (or Ge), which in crystalline form is a semiconductor but as a liquid is a metal. In the area of liquid metals and their alloys new efforts continue to focus on understanding electron transport, atomic transport, structure, and thermodynamics. Here the electrons may not be localized, but these systems are strong scatterers. The difficulties in understanding their static and frequency-dependent conductivities lies in our incomplete knowledge of the microscopic theory of dense classical liquids, which include the liquid metals (see Chapter 9). Some liquid binary alloys exhibit transport coefficients that actually become singular as a function of the relative concentration of the two species. Here the balance between electron-electron interactions and disorder can be altered by the alloying process. Consequently, the tendency toward localization can be increased and controlled by the experimentalist.

The range of the metallic state can be extended considerably both to low densities and to high temperatures. The resulting systems constitute forms of matter that are of intrinsic interest, because of both their similarity to dense, strongly coupled plasmas and their proximity to two quite fundamental phase transitions (liquid-vapor and metal-

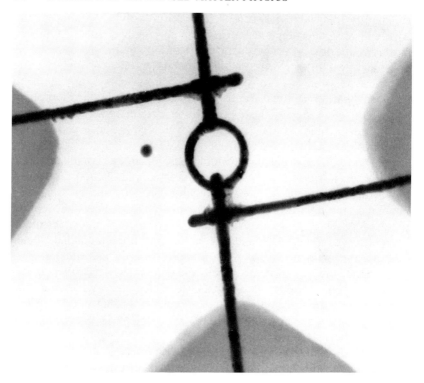

FIGURE 1.6   A gold metal ring of diameter approximately 3500 Å. The width of the line is approximately 400 Å. These rings are used to search for flux quanta of the kind seen in superconducting rings. The gold is not superconducting. The large darker area to which the lines attach are the pads that provide connection to the external world. The dot is used for calibration. (Courtesy of IBM Thomas J. Watson Research Center.)

insulator). On the vapor side, the interactions are largely short ranged; on the metal side, they are screened, long-ranged interactions. During the transitions, the character of the interactions changes, an unusual behavior in the context of the standard theories of critical phenomena and associated transport. Though the effects are extremely interesting and at the core of some fundamental issues, the experimental situation presents serious challenges because of the extremely high critical temperature of most metallic systems.

The conduction electrons play an important role in disordered metals, as they do in metallic crystals. They contribute not only to transport phenomena but, through screening, to the actual forces acting between the ions themselves. The forces are expressible in terms of two (and often higher) center potentials, which in turn determine the structure of the disordered metal. These interactions differ qualitatively from those in insulating systems. Because they vary with electron density, they can again be altered by alloying. This effect may be of some consequence in resolving the issue of why some metallic mixtures can be made to form metallic glasses, and why some can not.

MIXED MEDIA

It is now possible to fabricate materials that are mixtures of a number of constituents (either insulators or metals) and in which the characteristic length scales may be as small as 50 Å. Such heterogeneous, microcondensed forms of matter are particularly interesting because one can tailor desired bulk properties by altering the constituents, their size distributions, or their relative concentrations. As one example, the wavelength of ordinary light is a few thousand angstroms and generally exceeds such scales of inhomogeneity. But, so far as the optical properties of these systems are concerned, they appear to behave as continua, and now one can tune the basic dielectric properties in a way not often possible with homogeneous systems. It should be noted that mixed or granular media, or composites, often display a great deal of order at the microscopic scale yet should still be properly regarded as disordered systems. Depending on the disposition of the matter in these systems, the topology of the arrangements of the constituents, and their detailed connectivity, it is possible to find percolative and critical behavior characteristic of the localization problem (impurities in semiconductors, for instance) discussed above.

Some small metal particle systems show clear evidence of a superconducting transition persisting in the metal for particle sizes as small as 100 Å or smaller.

CONDENSED MATTER AT HIGH PRESSURE

By application of pressure to a sample, we change its density and hence its physical properties, often quite dramatically. However, condensed matter is generally considered quite incompressible, and to achieve even modest fractional changes in density has often required pressures of thousands or even tens of thousands of atmospheres.

During the past few years, notable advances in high-pressure physics have occurred. Though so-called dynamic techniques (shock-wave methods) have also developed, the advances in static high-pressure physics have centered largely on the active use of the diamond anvil cell (see Figure 1.7). With these devices it is possible to develop pressures in excess of 1 Mbar. More importantly, such pressures can change by a significant factor the average volume available to an atom or molecule in a solid or liquid. Minerals may be exposed to pressure ranges reminiscent of their original environment in the Earth's interior. This is having an impact on geophysics and planetary physics, as well as on materials science and solid-state chemistry. Though electrical and even thermal measurements are now possible in these devices, most of the probes used to detect changes induced in samples have been either optical or x ray in origin. They exploit the transparency of the diamond. It is possible to utilize diamond-cell devices in conjunction with radiation environments that are unusual in their degree of intensity, polarization, time structure, or wavelength (synchrotron radiation and pulsed lasers are examples of these).

One of the most striking uses of diamond cells has been the transformation of insulators into metals. This most fundamental of all phase transitions has been observed in molecular crystals, in ionic crystals, in transition-metal oxides, and in mixed valence compounds. For example, iodine has traditionally been regarded as an insulator, but it appears to be a metal at pressures in excess of 200 kbar. Above 1 Mbar the noble gas xenon should also become a metal, and at about 2 Mbar, even hydrogen should become metallic. Useful pressures much over 1 Mbar have not yet been achieved statically, however, but hydrogen has been compressed to about 7 times its normal low-

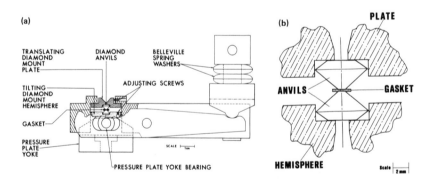

FIGURE 1.7   An ultrahigh-pressure diamond cell. The complete cutaway cross section (a) shows the essential components including the anvil supports, alignments design, lever-arm assembly, and spring-washer loading system. The detail (b) shows an enlargement of the opposed diamond anvil configuration with a metal gasket confining the sample. The cell was developed at the National Bureau of Standards. (Courtesy of G. Piermarini, National Bureau of Standards, Gaithersburg, Maryland.)

temperature solid density, and its vibrational modes have been studied using Raman-scattering techniques.

Though static methods are developing rapidly, advances in instrumentation and concept have also been reported in shock-wave physics. With these techniques a substance can be brought into high-temperature, high-pressure regions of its phase diagram that are inaccessible by any other means. These experiments yield information such as the compressibility, plasticity, phase stability, and optical properties of condensed matter pertinent to planetary and even stellar physics.

OPPORTUNITIES

Further simplifications in the pseudopotential and other band-theoretic techniques employed in the calculation of the electronic structure of perfect metallic crystals can be hoped for. Their application to defect- or disorder-related problems, however, will certainly hinge on the availability of substantially increased computational facilities. In view of the importance of this work to technology, further investment in it is certainly warranted.

It is expected that the density-functional method will be used widely in the future in the theoretical study of the electronic properties of crystalline solids and in cases, such as surfaces and crystalline defects, where the electron density is strongly inhomogeneous. Considerable effort will be devoted to trying to understand why this method, in its so-called local-approximation, works as well as it does, and why it occasionally fails.

Despite the considerable success of the preceding methods for the calculation of the ground-state, and even the excited-state electronic properties of metals, no comparably simple and accurate methods exist for the calculation of the excited electronic states of semiconductors and insulators, in which these states are separated from the ground state by an energy gap. These excited states are needed in the calculation of the response of such materials to time-dependent perturbations, such as externally applied electromagnetic fields. In view of the importance of being able to calculate such responses for the interpretation of data obtained by a variety of experimental probes, we can expect attention to be directed to the development of methods that will yield the excited states of semiconductors and insulators accurately.

High-pressure physics appears to be entering an exciting and productive phase and is a good example of a strong feedback mechanism operating between science and technology. High pressure is expected to play a prominent role in elucidating the physics of the metal-insulator transition, both in ordered and disordered systems. It will certainly be used to address the questions associated with s-d and f-d transitions in the heavy elements and also to aid in unraveling the puzzles of the interesting classes of intermediate-valence compounds and heavy-fermion systems. It may even shed light on the nature of the ground state of the light alkali metals, which have been thought of as reasonably well understood but continue to behave in ways (particularly at low temperatures) that are not easily explained. More generally, static high-pressure methods are expected to play an ever-increasing role in determining the electronic structure of new materials, in complex materials, in semiconductors, in artificial superlattice systems, in amorphous solids, in glasses both insulating and metallic, and in polymers, liquid crystals, and simple fluids and their mixtures.

Because of their continuing technological importance, disordered materials, including metallic glasses and amorphous semiconductors, are expected to receive growing experimental and theoretical attention in the future. Much has been learned in the past decade about the existence of new kinds of states, the so-called tunneling states, in highly disordered matter. However, much is still to be understood about the nature of the elementary excitations in glasses, especially at low temperatures. Glassy metals can be formed by rapid cooling techniques, during which they may possibly preserve certain aspects of the structure and dynamics of the liquid from which they were formed. Accordingly, such systems offer the prospect of studying the dynamics of disorder in a manner that is not possible when the system acts as a classical liquid. The glass transition (in insulating and metallic glasses) is not well understood, and further activity in this area, both experimental and theoretical, is expected. Crucial to this endeavor is a deeper understanding of the systematics of bonding in condensed matter within a framework going considerably beyond the current picture.

Though there has been a rejuvenation of mean-field theories used to describe the response properties of mixed media, they are still not well understood in detail. In particular, the far-infrared response of metal-particle composites is yet to be unraveled. Experimental probes, of both their real-space and electronic structure, will continue to be developed. The frequency dependence of their optical response in the superconducting state, and its ultimate understanding, will also be a subfield of interest.

Driven largely by the impetus toward very-large-scale integration in electronics, small structures can also be made with substantial long-range order (arrays of small metal spheroids, for example). The particles themselves are of such a length scale that many of the conventional methods of solid-state physics no longer apply in determining their essential physical properties. The field of ordered micro-condensed-matter science is still in its infancy but is perceived widely as one in which many of the traditional subareas of solid-state physics will yet have a considerable impact. In this technologically crucial area, interface states, ballistic transport, Kapitza resistance, quantum-well effects, electromigration and thermomigration, and noise are all topics of fundamental interest and will offer research opportunities for the future.

2

Structures and Vibrational Properties of Solids

INTRODUCTION

Matter in a solid state consists of many nuclei and electrons that form a structure in space. Knowledge of this structure is essential for understanding the physical properties of the solid, for example, whether it is a metal, a semiconductor, or an insulator or whether magnetic order can be produced by the electronic interactions. Vibrations of the nuclei around their average positions produce excited states of the solid structure. Since the nuclei have much heavier masses than the electrons, their characteristic vibrational frequencies, $\sim 10^{13}$ s^{-1}, are much lower than the frequencies of $\sim 10^{15}$ s^{-1} typical of many electronic excitations. These low-frequency vibrations are ubiquitous aspects of all solids: they propagate, and in so doing carry heat and information; they are important in the thermodynamics of solids; they are always present to absorb or scatter such experimental probes as electromagnetic radiation and neutrons, as well as other excitations in a solid such as electronic and magnetic excitations; and they lead to important electronic ordering effects such as superconductivity.

The vibrational properties of many solids can be understood on the basis of the harmonic approximation, in which the force acting on a given nucleus is assumed to be a linear function of the displacements of that nucleus and of the other nuclei from their average positions. The problem can be solved exactly in this approximation. The quantized

58

units of vibration are called *phonons*, and the quantum of energy is the vibrational frequency ω times Planck's constant $\hbar$. The nature of the vibrations is closely related to the structure of the solid: in crystals they form collective propagating excitations whose frequencies are described by dispersion curves $\omega = \omega(\mathbf{k})$, where $\mathbf{k}$ is the wave vector associated with the excitation (the direction of $\mathbf{k}$ is the direction of propagation of the excitation, and the magnitude of $\mathbf{k}$ is $2\pi/\lambda$, where λ is the wavelength of the excitation); in disordered systems they are sensitive to the topology of the structure of the solid and to the local order. The addition of small anharmonic forces leads to finite lifetimes and scattering of these phonons. In some systems the anharmonic forces are large, however. The dynamics of the nuclei in strongly anharmonic systems may be qualitatively different from the behavior of simple oscillators. There can be stable nonlinear excitations termed solitons, interesting statistical mechanics of thermally excited interacting vibrational states, and phase transitions to structures of different symmetry.

Because this field is extensive and closely related to other topics, many of its aspects are considered in separate chapters, in particular, critical phenomena at phase transitions, structures of surfaces and interfaces, defects in crystals, and properties of particular classes of solids.

THEORETICAL CALCULATIONS

The primary goal in the theory of the structures of solids is to understand both why different types of solids form and how the resultant structures control the properties of solids. This is a many-body problem involving $\sim 10^{23}$ electrons and nuclei. One of the highlights of research during the past decade is the progress toward a unified theoretical understanding of the combined many-electron/many-nucleus problem. Indeed, predictions of the structures and vibrational excitations of solids are currently a crucial test of our understanding of the ground state of the electronic system.

Since the mid-1970s, there has been a qualitative change in the ability to predict structures and related properties of solids *a priori* without using any information from experiments. This rapid development has been made possible both by the increase in power and availability of computers and by the formulation of new ways to treat the quantum many-body problems. Of these developments in the treatment of the electronic system discussed in Chapter 1, it is the density functional approach to electronic exchange and correlation that is the basis for the recent progress in accurate first-principles calculations of a wide range of structural and vibrational properties of solids. Other techniques, such as the many-body Monte Carlo quantum methods, make it possible to study the simplest cases in great depth.

Among the primary achievements of such calculations are the phase diagrams of many elements and compounds as functions of pressure. Recent results include the structures of transition metals, semiconductor-metal transitions, graphite and diamond structures of carbon, and many other crystals. An exemplary case is hydrogen, which is expected to transform from an insulating molecular system to a metallic solid at high pressure. This is illustrated in Figure 2.1, which gives the total energy versus average proton separation, found from quantum Monte Carlo calculations. Similar results are found from perturbation

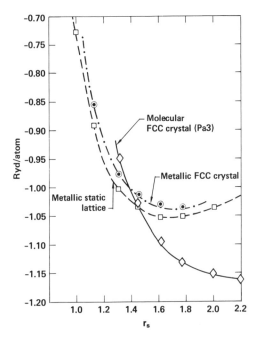

FIGURE 2.1   Ground-state energy of hydrogen as a function of the average proton separation a, in units of the Bohr radius a_0 calculated by an approximate Monte Carlo simulation of the many-body fermion problem. The solid curve gives the energies for molecular and monatomic metallic phases. The dashed curves show the effect of fixing the protons, i.e., eliminating their zero-point motion, in the metallic phase. The results indicate a first-order transition from the molecular to the metallic phase near the crossing point at $a \sim 1.35a_0$. (Courtesy of D. M. Ceperly and B. J. Adler, Lawrence Laboratory.)

theory and density functional calculations that have also considered different metallic structures. The salient result is that hydrogen will become a metal at pressures $\gtrsim 2$ mbar, which may be achieved in diamond anvil cells in the near future.

One of the currently interesting developments is the emergence of a unified theoretical approach to structures, vibrations, and electron-phonon interactions. This is made possible by density functional calculations for crystals with atoms displaced from their equilibrium positions to determine small energy differences, forces acting on individual atoms, and the macroscopic stress. From the restoring forces and stresses, the vibrational properties can be obtained with no input from experiment. Calculations to date include complete phonon dispersion curves $\omega(\mathbf{k})$, the pressure dependence of phonon frequencies and other anharmonic coefficients, and anomalous soft phonon modes. Results of calculations carried out so far agree with experiments to within a few percent and predict other properties not known experimentally.

MEASUREMENTS OF STRUCTURES AND PHONON SPECTRA

The basis of experimental measurements of the structures and dynamics of condensed matter is the absorption or the scattering of particles whose momentum and energy can be measured. The average structure is measured by the intensity of scattering as a function of the difference between the momenta of the incoming and outgoing particles. Dynamical information can be obtained by measurement of the energy lost or gained by the particles. Conservation of energy requires that the excitation that is created (or destroyed) has energy equal to either (1) the energy of a particle that is absorbed (or emitted) or (2) the difference between the incoming and outgoing energies of a particle that is scattered inelastically. Experimental probes used in current investigations of structures and dynamics include x rays, photons, neutrons, electrons, atoms, and ions. Experiments using electrons and atoms are particularly suited for studies of surfaces and are discussed in Chapter 7.

Inelastic neutron scattering is a powerful technique for the study of the dynamics of atoms in condensed matter (see Appendix E). The spectrometers needed to resolve the energies of the neutrons, developed in the 1950s and 1960s, continue to provide an extensive body of knowledge on phonons and other excitations in condensed matter. Recently neutron scattering has provided the crucial short-wavelength probe for exploration of the challenging problems associated with phase transitions, anomalous phonon dispersion curves due to strong electron-phonon interactions, and dynamics of nonlinear systems.

There have been two major advances in neutron-scattering methods recently. One is the development of a neutron spin-echo spectrometer, which can measure energy transfers as small as a few microelectron volts. This resolution makes it possible to determine the dynamics of low-frequency, quasi-elastic phonons and the intrinsic lifetimes of phonons. The second is the advent of spallation sources, which are described in Appendix E. These sources produce neutrons with large usable ranges of momentum and energy that can provide increased spatial resolution and measurements of high-energy phonons, particularly those involving light atoms such as hydrogen.

Experiments using x-ray scattering and absorption have become much more powerful because of recent advances in the production of intense, tunable x rays from synchrotron sources. The increase in angular resolution and intensity has made possible new experiments. One is the study of melting of two-dimensional systems of rare-gas atoms, described in Chapter 3. Another is the first measurement of the phase of the scattered x rays. This advance offers the possibility of yielding powerful new information on structures but is controversial at present. The pulsed nature of synchrotron radiation has been utilized to measure the rapid melting and recrystallization on nanosecond time scales that occur in pulsed-laser annealing. There has been an enormous increase in the number of measurements of extended x-ray absorption fine structure (EXAFS) spectra, which are being used to determine the local environment of a given type of atom. The most important results have been obtained for alloys, disordered solids, ionic conductors, and liquids, where EXAFS provides detailed information on the correlation functions of different atoms.

The interaction of light with solids provides many of the most useful and versatile techniques for studying the dynamics of condensed matter. Although the range of momenta that can be studied by this technique is limited compared with that of neutron scattering, the absorption and scattering of photons have much greater resolution, dynamic range, and sensitivity than is possible with neutron scattering. Furthermore, because light couples to phonons primarily through the electronic polarizability, these experiments provide unique information on linear and nonlinear interactions of electrons, photons, and phonons.

The modern era of light scattering began in 1962 when lasers were first introduced as monochromatic sources of light. Since that time, Raman scattering has become the most widely applied technique to determine vibrations in solids. In recent work, for example, scattering from tiny crystals under the extreme pressures that are generated in diamond anvil pressure cells is giving much new information on the nature of matter at compressed density. The use of optical interference enhancements has made possible detection of the vibrational spectra of molecules adsorbed on surfaces at submonolayer densities and of crystalline compounds formed in very thin ($\sim$20 Å) layers at interfaces between different solids. In addition, the use of intense laser beams and optical nonlinearities leads to new effects, such as coherent stimulated Raman scattering and hyper-Raman scattering involving several photons. The former has made possible lasing at Raman frequencies in optical fibers. The latter leads to different selection rules, so that vibrations can be detected that are not observable by ordinary Raman scattering.

Inelastic scattering of light with small frequency shifts $<10^{10}$ Hz, often termed Brillouin scattering, has expanded greatly, aided by development of highly selective multiple-pass interferometers. Among the recent accomplishments of this technique are measurements of acoustic vibrations in metals through inelastic reflection caused by dynamical rippling of the surface. Low-frequency scattering also plays a crucial role in investigations of nonlinear systems, including such problems as the detection of tunneling modes in glasses, ionic motion in superionic conductors, large increases in quasi-elastic scattering near phase transitions, and dynamics of incommensurate structures described later.

The nature of the coupling of electrons and phonons can be studied by resonance scattering, in which selected electronic states are enhanced by their resonance with the light frequency. Because the extreme resonance conditions occur at energies where the light is absorbed, understanding the phenomenon has required the development of theoretical tools to deal with the difficult problems of nonequilibrium excited states coupled to the stochastically fluctuating environment. This has been applied particularly to investigate impurity states coupled to the lattice and the scattering mechanisms for electrons and holes in semiconductors.

Infrared light can be used to study optically active phonons through reflectivity and absorption. As in the scattering experiments, the advent of infrared lasers has made possible new experimental areas, and many recent advances have been in the areas of low-frequency measurements. This is one of the powerful tools for studying ionic conductors, amorphous metals, and the coupled electron-phonon system in semiconductors.

Each of these experimental tools for determining structures and dynamics has an important role in exploring the properties of solids and the physics of condensed matter. Some of the highlights and opportunities made possible by these techniques are mentioned below.

PHONON TRANSPORT

In the area of phonon transport varied aspects of phonons as elementary excitations of condensed matter are explored: the spectrum of energies, the velocities of propagation, scattering and decay of phonons, and their interactions with defects and other excitations. Before 1965, phonon transport was almost always studied by measuring the temperature dependence of the thermal conductivity. This yields a transport coefficient that is an average over different scattering processes due to anharmonicity, defects, and surfaces, weighted by the equilibrium distribution of phonons. In contrast, new techniques for generation and detection of high-frequency phonons have made possible the direct study of phonon properties, selected by their frequency, velocity, and direction of propagation, in frequency ranges extending to >1 THz (10^{12} s^{-1}).

The initial experiments used heat pulses and measurement of the time of flight of phonons from heater to detector. They could resolve individual phonon modes, which propagate ballistically with their respective group velocities, as well as diffusive heat transport resulting from multiply scattered phonons. Important results included the observations of second sound, the propagation of temperature waves in solids, and the propagation of solitons. The latter are well-defined excitations of a nonlinear lattice. This work was a stimulus for interest in nonlinear problems in other areas.

There are several new methods of energy-selective generation and detection of high-frequency phonons. These include phonon-assisted tunneling, optical techniques, and time-of-flight selection of high-frequency phonons using the dispersion of velocities. Superconducting tunnel junctions bonded to the sample surface can selectively study phonons with energies up to the superconducting gap of ~0.5 THz. Optical techniques utilizing visible lasers can be used in many transparent solids to generate and detect phonons through coupling to sharp impurity states. State-of-the-art techniques of pulsing and focusing visible lasers make possible complete studies with simultaneous spectral, spatial, and temporal resolution. Also, phonons can be generated by infrared lasers using surface piezoelectric effects. This approach has the potential of creating phonons with phase coherence limited only by surface roughness.

Transport of energy in different phonon modes has been shown to vary enormously. In particular, low-frequency transverse phonons can often propagate over large distances, and their weak scattering mechanisms can be studied in detail. Perhaps the most dramatic experimental consequences of the long lifetime of certain acoustic phonons are the phenomena of phonon imaging and focusing caused by anisotropy in the velocity of propagation. For example, phonons produced by a heater at one point on a sample can be focused along particular crystallographic directions and can propagate ballistically for distances of ~1 cm under readily achievable conditions. An example of this striking anisotropic transport of energy in germanium is shown in Figure 2.2.

Other developments include the study of anisotropic phonon winds and their effect on the shape of electron-hole droplets in semiconductors; measurement of the frequency dependence of scattering by defects such as donors and acceptors in semiconductors; stimulated directional emission of phonons; demonstration of phonon mirrors created by superlattices of semiconductors; measurement of lifetimes of optic phonons in the picosecond range; generation and study of high-frequency surface phonons; and observation of anomalous transport in glasses at low temperatures due to coupling to low-frequency tunneling modes.

ELECTRON-PHONON INTERACTIONS

The interactions of phonons with photons, electrons, magnons, and excitons are indispensable ingredients in understanding the physical properties of solids. In cases of weak coupling, the phonons cause scattering, which is an important limitation on the mean free path of electrons, i.e., on the conductivity of metals and the mobility of carriers in semiconductors. Since the electrons also affect the phonon frequencies, the same interactions can be manifested in anomalous dispersion of the phonon frequencies and in phase transitions such as superconductivity and structural transitions. There can also be nonlinear solutions for localized electronic states coupled to atomic displacements. The best known recent example is the formation of fractionally charged solitons in conducting polymers (Chapter 10). For reasons such as these, electron-phonon interactions are of great importance in solid-state physics, and there is a growing interest in studying and utilizing the consequences of these interactions.

The transition metals and their compounds are the focus of much of the activity in this area because the electron-phonon interactions are

FIGURE 2.2   Phonon focusing. The bright areas represent heat energy propagating to the surfaces of a germanium crystal produced by a pulse of heat at a point on the back surface of the crystal. The phenomenon is caused by intense channeling of heat flux along certain crystal directions. (Courtesy of J. P. Wolfe, University of Illinois.)

thought to be responsible for high-temperature superconductivity in compounds like V_3Si and NbC, as well as for phonon softening and displacive phase transitions. A striking example of experimental and theoretical work is the ω-phase transition, in which the bcc structure is unstable to displacements of planes of atoms perpendicular to the (111) axis. The dynamics of this transition in Zr have been studied by neutron scattering, which has detected an anomalously low phonon frequency shown in Figure 2.3 and an increase in intensity of the central-peak scattering at zero frequency at the wavelength corresponding to the periodicity of the ω-phase. Theoretical density-functional calculations have determined an entire curve for the energy as a function of the positions of the planes, giving the low phonon frequency, two stable solutions in the bcc and ω-phase structures for Zr, and insight into why the effects are greatly reduced in the neighboring elements Nb and Mo.

The electron-phonon interactions in transition-metal compounds have also made possible a new class of experiments involving light scattering, normally not observable in metals. The same interactions that cause the phonon anomalies also give rise to coupling to the light through the electrons. For example, $NbSe_2$ distorts into an incommensurate structure (discussed below) owing to the electron-phonon coupling, and the dynamics of the atomic displacements have been

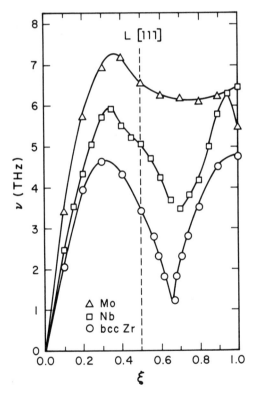

FIGURE 2.3   Phonon dispersion curves for the longitudinal (111) branch measured by inelastic neutron experiments on Mo, Nb, and the high-temperature (1400 K) bcc phase of Zr near the ω-phase transition. (Courtesy of C. Stassis and B. N. Harmon, Iowa State University.)

detected in the light-scattering spectrum presented in Figure 2.4. Perhaps the most striking observation is the new peaks at low temperatures, interpreted as electronic excitations across the superconducting gap. These results have led to new theoretical and experimental work to understand the basic phenomena involved and the role of the interactions in superconductivity and other properties.

Other areas in which electron-phonon interactions play a crucial role are inelastic electron tunneling and a new experimental technique termed point-contact spectroscopy. The use of tunneling spectroscopy in superconductors to determine phonon densities of states, weighted by electron-phonon couplings, is now well established. Recent advances in making tunnel junctions of superior quality have made possible tunneling in transition metals, high-temperature superconductors, and magnetic superconductors. Such measurements on magnetic superconductors show the disappearance of the superconducting energy gap as the magnetic transition is approached. In the high-temperature superconductors, e.g., Nb_3Sn, tunneling results indicate

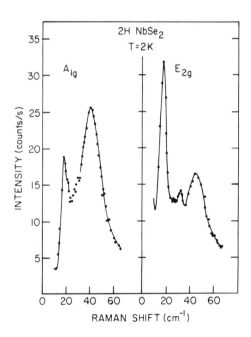

FIGURE 2.4   Raman spectrum of 2H $NbSe_2$ at low temperature. The peaks at ~40 cm^{-1} are amplitude modes of the incommensurate structure and those at ~20 cm^{-1} are excitations of the electrons across the superconducting gap. (Courtesy of M. V. Klein, University of Illinois.)

that, of all the phonons, those of low frequency are most effective in promoting superconductivity.

Point-contact spectroscopy involves measuring the current-voltage relation for a current of electrons through a metallic point. If the dimensions of the point are smaller than the electronic mean free path, electrons can be accelerated to the energy eV, where V is the voltage drop. Measurement of the current as a function of V gives direct information on the energy dependence of the scattering mechanisms. At present, theoretical work is attempting to derive the relations to the underlying phonon properties. One advantage of this technique is that it can be applied to many materials and is not restricted to superconductors.

DISORDERED SOLIDS AND INCOMMENSURATE PHASES

A growing area of research is concerned with disordered solids that present intellectual challenges, unique phenomena, and extensive applications. One class of disordered materials is the amorphous or glassy solids, which have no long-range order. The atomic structures of glasses, nevertheless, have characteristic types of short-range order, e.g., favored coordination numbers and angular arrangements of the nearest neighbors associated with specific types of bonding. For example, in vitreous silica the oxygen atoms have twofold coordination and the silicon atoms have fourfold tetrahedral coordination, whereas in amorphous metals the coordination number is higher, ~8-12. Experimental information on the short-range order is obtained by diffraction of x rays, neutrons, and electrons and by EXAFS, which determine angle-averaged radial distributions of the probability of finding neighboring atoms. These measurements cannot determine the three-dimensional structure uniquely, but they provide stringent conditions on models of the structure. Research in this area has increased dramatically in recent years owing to the availability of synchrotron facilities as intense, tunable, collimated x-ray sources and the advent of spallation facilities as sources of higher-energy neutrons, which can give improved spatial resolution.

The intellectual challenges that have highlighted recent research in this area are concerned with the ways that groups of atoms with short-range order can be connected together to build space-filling rigid structures with no long-range order. An interesting contribution to the theory of such structures is the demonstration that small sets of regular polyhedra can be packed to generate nonperiodic, disordered structures that fill three-dimensional space. There are, however, many degrees of freedom to consider in a physical glass, and there is much controversy and continuing research on the thermodynamics of the glass transition and the nature of the structures formed. The vibrational excitations are especially pertinent to the studies of disordered structures because they depend sensitively on both the short-range order and the connectivity or topology of the structure. Theoretical studies of vibrational properties of strongly coupled disordered networks, especially with topological disorder, have led to new perspectives on excitations in disordered systems. Experimental measurements of vibration frequencies in glasses, together with the improved theoretical understanding, have motivated new explorations of the topology of glasses, such as silica.

Another aspect of the dynamics is the existence of low-frequency modes, which appear to occur universally in disordered systems. These nonlinear excitations dominate many low-frequency aspects of glasses, e.g., low-temperature heat capacity, thermal transport, electrical resistivity, and dielectric loss. Although they are thought to involve finite displacements of atoms by tunneling or thermal hopping, the microscopic origins of these modes are unknown.

A different class of disordered solids are crystals in which there is intrinsic disorder. The two areas of most current interest are ionic conductors and plastic crystals. Crystals called superionic conductors contain large densities of ions that can diffuse with rates comparable with those of ions in liquids. For example, in the high-temperature phase of AgI the I ions form a solid bcc lattice in which the Ag ions are as mobile as in the melt. Studies of these materials have been stimulated by their technological applications. The term plastic crystal denotes crystals containing molecules that are orientationally disordered. The low-frequency reorientations that these molecules can undergo are strongly coupled anharmonic motions, which lead to unusual mechanical properties of these solids. For ionic conductors, plastic crystals, and other dynamically disordered systems, the basic questions are: Why do such crystals form, and how do the ions or molecules move? Investigations on a microscopic scale currently utilize x-ray and neutron scattering, EXAFS, nuclear magnetic resonance, light scattering, high-frequency conductivity, and theoretical work on these highly anharmonic, nonlinear problems.

An exciting class of structures is one in which there are simultaneously two incommensurate periodicities coexisting in the same solid. Such a structure is not periodic because there is no translation that is equal to integral numbers of primitive translations of both periodicities. However, each periodicity can be separately observed in a scattering experiment. Such structures were known for some time (e.g., the spin density wave in chromium), but only in the last decade have they taken their place in the field of phase transitions and their symmetries and dynamics studied extensively.

Several types of incommensurate solids have been found. In one, which has been discussed in Chapter 1, the electron-phonon interaction stabilizes a distortion with the Fermi wave vector k_F, which is incommensurate with the lattice periodicity. Examples include chain compounds like TTF-TCNQ and layered metals like $NbSe_2$. A different mechanism that can occur in either metals or insulators is a zero phonon frequency at an incommensurate wave vector $\mathbf{k}$, which can be caused by simple combinations of interatomic forces. This is a soft mode that leads to a phase transition, as happens in K_2SeO_4 and $ThBr_4$. Another type of incommensurate structure results from the coexistence of interpenetrating lattices with different periodicities. An example is $Hg_{2.72}(AsF_6)$ in which the mercury atoms form linear metallic chains with an average spacing that is incommensurate with that of the AsF_6 lattice.

The vibrational states of incommensurate systems differ from those of ordinary crystals in fascinating ways. In particular, since it requires no energy to slide or change uniformly the relative phase of one periodicity relative to the other, there may be phason excitations with zero frequency at infinite wavelength. In the harmonic approximation, there is a phason dispersion curve with frequency linear in wave vector $\mathbf{k}$ at small $\mathbf{k}$, in addition to the ordinary sound modes present in all solids. There has been a widespread search for these modes leading to their observation in $ThBr_4$ and $Hg_{2.72}(AsF_6)$ by neutron scattering. The difficulty in observing these modes at longer wavelength and lower frequency, e.g., in light-scattering experiments, appears now to be understood in terms of a fundamental difference between phasons and true acoustic modes. The latter become more precisely defined propagating modes as the frequency decreases, whereas the phasons are greatly modified by anharmonicity and become overdamped at low frequencies. The strongly nonlinear character of phason modes leads to domainlike descriptions of incommensurate phase transitions like those described below.

PHASE TRANSITIONS AND NONLINEAR EXCITATIONS

Phase transitions that involve a change in the structure of a solid are among the archetypal examples of this general phenomenon. There are two paradigms for structural transitions—order-disorder and displacive. The former is a change in degree of disorder present in the structure. The latter involves displacement of atoms from sites of high symmetry to ones of lower symmetry. Each paradigm is illustrated in the previous two sections by recent work on disordered crystals, incommensurate phases, and structural transitions, such as the ω transition in Zr. These and other phase transitions, e.g., ferroelectricity, continue to provide major conceptual challenges and phenomena with technological applications.

Research on nonlinear excitations involving finite displacements of atoms has become a stimulating area of physics. Although exact solutions to simple nonlinear models and phenomena like solitary waves have been known for many years, a veritable explosion in the study of such excitations has occurred in condensed-matter physics since the mid-1970s. An impetus to this work was the progress in understanding displacive phase transitions, where studies of the dynamics revealed domain wall-type solutions that cannot be represented by perturbation expansions in the displacements of the atoms from their equilibrium positions. The dynamics of such systems consist not only of spatially extended, small-amplitude phonons, but also of spatially compact, large-amplitude excitations, often referred to as solitons. Although this has developed into an exciting new subfield, there is still controversy over how these excitations affect the thermodynamics of phase transitions.

Many stimulating developments in nonlinear dynamics have been made in the context of quasi-one-dimensional systems. A particularly interesting case is the conducting polymer polyacetylene $(CH)_x$, whose properties are striking consequences of the electron-phonon interaction. They are described in detail in Chapter 10. The general ideas underlying such excitations have widespread ramifications in physics and are discussed in Chapters 1, 3, 4, and 11.

OPPORTUNITIES

The ability to carry out theoretical calculations that predict the structures and vibrational properties of solids is expanding rapidly and will play a major role in future work. Because the calculations can be done accurately for real solids, there is emerging a new relation between theory and experiment and a more unified understanding of structural, vibrational, and electronic properties of matter. The potential of future work is to develop new ideas and methods for excited states and nonzero temperatures, to make simple models that describe the essential points, and to gain greater insight into the nature of condensed matter.

New experiments on structures and dynamics of solids can be made possible by improved synchrotron sources of x rays and by high-flux steady-state or pulsed sources of neutrons. Exciting possibilities include direct determination of structures using the phases of scattered x rays, measurements of fast transient structures, and improved energy resolution that can enable inelastic scattering of x rays to measure dynamics of atoms and electrons. High fluxes of neutrons would enable measurements to be made with greater resolution and on the small samples often crucial for forefront research. Pulsed spallation sources will permit inelastic scattering at high energies, e.g., at energies comparable with those of the vibrations of hydrogen atoms.

Current and future innovations in light scattering, such as femtosecond pulses and resolution of small frequency shifts, will make possible experiments on new materials, conditions, and time scales. Important contributions will likely occur for fast-reaction kinetics, properties of surfaces and interfaces, phase transitions, nonlinear excitations, and novel superconductors, for example.

High pressures achievable in diamond anvil cells open many possibilities for understanding why structures form and creating states of matter never before accomplished in a laboratory, such as metallic hydrogen.

Future areas of research in phonon transport will likely include increased emphasis on lower-dimensional systems, superlattices, nonlinear lattices, transport of phonons through interfaces, phonon dynamics in the subpicosecond range, and coherent excitations. The most important need for future work is the development of simple, sensitive tunable generators and detectors of phonons to extend measurements to wider classes of materials. This work will also have an impact on other areas of condensed-matter science, such as heat transport in small fast electronic devices, transfer of energy in pulsed-laser annealing, and steps toward development of a phonon laser.

The structures of glasses and other solids with disorder are at present only partially understood, and there is much controversy concerning the degree to which spatial order extends to intermediate ranges. New information and ideas are needed to understand such basic features of the structures of disordered solids. The microscopic origins of the low-frequency modes that occur almost universally in disordered systems are also unknown. Investigation of these modes by many different techniques will be an important area of future research in disordered systems.

The theoretical and experimental study of solitons and other nonlinear phenomena is an exciting area of research with many fundamental questions to be answered, such as the stability of solitons to small displacements, their role in phase transitions, effects of quantum fluctuations on them, and the nature of fractionally charged excitations. There are many possibilities for entirely new nonlinear phenomena in physical systems that may be realized through imaginative ideas and novel synthesis of materials.

Synthesis of new materials will likely provide unforeseen structures and phenomena as stimulating as those of the recent past, such as organic conductors and superconductors, incommensurate structures, and lower-dimensional systems. The creation of man-made artificial structures, such as semiconductor superlattices, is just beginning to reveal the range of new possibilities. Studies of structure and vibrations will certainly continue to probe phenomena of intrinsic interest as well as to provide keys to understanding the nature of new materials.

3

Critical Phenomena and Phase Transitions

INTRODUCTION

One of the most active areas of physics in the last decade has been the subject of critical phenomena. Enormous progress was made during the decade, both theoretically and experimentally, and research in the field was honored with the award of the 1982 Nobel prize in physics. It seems safe to predict that the study of critical phenomena and closely related subjects will remain a major activity of condensed-matter physics throughout the 1980s and that much further progress will occur.

WHAT ARE CRITICAL PHENOMENA, AND WHY ARE THEY INTERESTING TO PHYSICISTS?

The term critical phenomena refers to the peculiar behavior of a substance when it is at or near the point of a continuous-phase transition, or the critical point. A continuous-phase transition, in turn, may be defined as a point at which a substance changes from one state to another without a discontinuity or jump in its density, its internal energy, its magnetization, or similar properties. The critical point or continuous-phase transition may be contrasted with the more familiar case of a first-order phase transition, where the above-mentioned properties do jump discontinuously as the temperature or pressure passes through the transition point. Continuous-phase transitions in many cases, but not all, are associated with a change of symmetry of the system.

Although the critical point was first discovered more than 100 years ago, a good understanding of behavior near a critical point has only emerged recently. The peculiarities of the critical point arise because there are, in each case, certain degrees of freedom of the system that show anomalously large fluctuations on a long-wavelength scale, compared with those of a normal substance far from a critical point. These large fluctuations cause a breakdown of the normal macroscopic laws of condensed-matter systems, in some dramatic ways and in some subtle ways, and it has been a major challenge to learn what are the new special laws that describe the systems at their critical points. The challenge has been difficult for theorists because the large fluctuations could not be handled by the old calculational schemes, which depended implicitly on long-wavelength thermal fluctuations being small. The challenge has been difficult for experimentalists, because in order to make measurements sufficiently close to a critical point, to test existing theoretical calculations, or to discover directly the laws of critical behavior where no theory exists, it is necessary to have extremely precise control over the sample temperature, and frequently over the pressure and purity as well.

The study of critical phenomena has been rewarding in spite of its difficulties, and the understanding gained has proved useful to the understanding of other types of systems—including quantum field theories in elementary-particle physics, analyses of phenomena in long polymer chains, and the description of percolation in macroscopically inhomogeneous systems—in which fluctuations play an important and subtle role but where precise direct experiments may be even more difficult than in the case of critical phenomena. The techniques of renormalization-group analysis, developed in the theory of critical phenomena, have had a profound impact on an entire branch of mathematics, for example in the study of iterative maps, which has applications to economics, biology, and other sciences, as well as to the study of nonlinear fluid dynamics and other problems in condensed-matter physics.

Experimental research on critical phenomena has also had an impact both inside and outside condensed-matter physics. The requirements of experiments on critical phenomena have often stimulated the synthesis of samples with a new degree of perfection and of materials with special properties—such as magnetic systems with anisotropic spin interactions. The precision measurement techniques developed for the study of critical phenomena have also found application, for example, in the study of the onset of fluid convection.

In the following sections, we further define the features of a critical point, and we give some examples of properties that show critical-point anomalies. We outline the progress that has been made in the field, and we give a few selected examples of important problems that are still unsolved.

EXAMPLES OF PHASE TRANSITIONS AND CRITICAL POINTS

Several examples may illustrate the difference between a first-order transition and a continuous transition or critical point.

One example of a critical point is the Curie point of a ferromagnetic substance such as iron (T_c = 770°C for iron). At temperatures below T_c, a single-domain sample of iron has a net magnetization **M** that points arbitrarily along one of several directions that are energetically equivalent, in the absence of an external orienting magnetic field. The strength of the magnetization $M(T)$ decreases with increasing temperature until the Curie temperature T_c is reached. Above T_c the magnetization is zero in the absence of an applied magnetic field, and we say the material is in a paramagnetic state. In most magnetic systems (including iron) the magnetization $M(T)$ decreases continuously to zero as the temperature approaches T_c from below; then we say that there is a continuous phase transition, or critical point, at T_c. In some cases, however, the magnetization of a substance approaches a finite, non-zero value, as T approaches T_c from below, and the magnetization jumps discontinuously to zero, as the temperature passes through T_c. In these cases there is a first-order transition at T_c.

In the magnetic example there is a symmetry difference between the phases involved, since the ferromagnetic phase has a lower symmetry than the paramagnetic phase.

The familiar boiling transition, from liquid to vapor, is a first-order transition. Thus, when water boils at 1 atmosphere pressure and a temperature of 100°C, there is a decrease in density by a factor of 1700. However, if the pressure is increased, the boiling temperature increases, and the difference in density between the liquid and vapor becomes smaller. There exists a critical pressure P_c where the density difference between liquid and vapor becomes zero; at this pressure the phase transition is no longer first order, and the transition temperature T_c at pressure P_c is described as the gas-liquid critical point of the substance. For pressures greater than P_c there is no distinction at all between liquid and vapor. We may note that there is no symmetry difference between the liquid and vapor phases.

In order to facilitate the comparison between critical points in various systems, it has proved convenient to introduce the concept of an order parameter associated with each phase transition. For systems like the ferromagnet, where there is a broken symmetry below T_c, the order parameter is a quantity like the magnetization, which measures the amount of broken symmetry in the system. For systems without broken symmetry, one chooses some quantity that is sensitive to the difference between the two phases below the critical temperature and measures the difference of this quantity from its value at the critical point. For the liquid-vapor critical point, we may choose the order parameter as the difference between the actual density of the fluid and the density precisely at the critical point.

HISTORY

The earliest theories of critical phenomena, developed near the end of the last century and at the beginning of this century, gave a good qualitative description of the behavior of a system near its critical point. However, it gradually became clear in the mid-twentieth century that these classical theories were incorrect in important details.

A most important step in this realization occurred in the 1940s, when Onsager found a remarkable exact solution of a model of a magnetic system in two dimensions (known as the two-dimensional, or 2-D, Ising model) and showed that its phase transition did not follow all the predictions of the classical theories. In the 1960s, experiments on actual three-dimensional (3-D) systems began to show more and more clearly that their critical behavior was also different from that predicted by the classical theories and different from that of the 2-D Ising model as well. At the same time, there appeared a certain regularity to the behavior of different 3-D systems, which was encouraging to the search for some general theory of these transitions. Other evidence for this viewpoint, and hints at the shape that the new theory must take, were provided by various types of numerical calculations (one might call them computer experiments), which included both computer simulations of thermal fluctuations in simple magnetic models and also numerical extrapolations of the properties of these magnetic systems from temperatures far above the critical temperature, where accurate calculations could be done.

An important step forward in our understanding of critical phenomena occurred in the mid-1960s, with the development of a set of empirical scaling laws, which were successful in describing certain relations between different critical properties of a system, although they could not predict all these properties from the beginning. The concept of universality classes developed, as it appeared that systems could be divided into certain broad classes, such that all members of a given class had identical critical properties but that these same properties varied from one class to another. One important factor that affects the critical behavior is the spatial dimensionality of the system—e.g., 3-D systems have different critical behavior than 2-D systems—but there are other factors that are relevant, including the symmetry differences between the states at the phase transition, the presence or absence of certain long-range interactions, and other factors that will be discussed below. A proper understanding of the factors that determine the universality class of a system had to await the developments of the 1970s, however, and in fact, a classification in the more difficult cases remains one of the tasks for the 1980s.

The most important theoretical advance of the 1970s was the development of a set of mathematical methods known as renormalization group techniques. These methods are not limited to critical phenomena—they are useful whenever one has to deal with fluctuations that occur simultaneously over a large range of length scales (or energy scales or time scales, for example). The methods proceed by stages, in which one successively discards the remaining shortest-wavelength fluctuations until only a few macroscopic degrees of freedom remain. The effects of the short-wavelength fluctuations are taken into account (approximately) at each stage by a renormalization of the interactions among the remaining long-wavelength modes.

The renormalization group techniques have made possible a number of achievements.

1. They have given us a justification for the scaling laws of the 1960s.
2. The renormalization group methods enable one to predict with high reliability which features of a system are relevant to determining its universality class for critical behavior and which features of the microscopic description become irrelevant in the vicinity of the critical point.
3. The renormalization group methods enable us to calculate properties of any given universality class. In the simpler cases, these critical properties have been calculated with a high degree of accuracy; and these predictions of the renormalization group have been confirmed in turn by some beautiful experiments of high precision. In more compli-

cated cases, the numerical accuracy of existing renormalization group calculations is not high, and further improvements in them are badly needed.

WHAT DOES ONE MEASURE?

In order to make more precise our discussion of critical phenomena, it is useful to give some examples of quantities measured and to give some examples of the laws that describe them.

Perhaps the most fundamental measurement to make in the vicinity of a critical point is to determine the way in which the magnitude of the order parameter approaches zero, as the critical point is approached from the low-temperature side. According to the classical theories of phase transitions, such as the van der Waals or mean-field theories, the order parameter should approach zero as the square root of the temperature difference from T_c. We may write this as

$$M = M_0(T_c - T)^\beta, \tag{1}$$

where M is the order parameter on the coexistence curve (i.e., for a ferromagnet, M is the magnetization in zero magnetic field; near the gas-liquid critical point M is proportional to the density discontinuity between liquid and vapor). M_0 is a constant that will vary from one system to another, and the exponent β is equal to 1/2 for all critical points, in the classical theories. Now the result of the modern theory of critical phenomena is that the classical theory is not correct close to T_c. We can still write the temperature dependence of the order parameter in the form of Eq. (1), but the value of the exponent β is not equal to 1/2. For the 2-D Ising model of magnetism, and for other 2-D systems in the same universality class, the result is $\beta = 1/8$, as given by the Onsager solution. For the gas-liquid critical point in three dimensions, as well as for the 3-D version of the Ising model, the result of the most accurate experiments and renormalization group calculations is $\beta = 0.325$, with an estimated uncertainty of ± 0.001. Other 3-D systems may belong to different universality classes, but their values of β are typically in the range 0.3-0.4.

The forms of the power law Eq. (1), for various values of the exponent β, are illustrated by the curves in Figure 3.1. The curves for $\beta = 1/8$, 0.325, and 1/2 are all qualitatively similar, and indeed the quantitative differences appear small on this linear scale. The differences may actually be quite large, however, if precision measurements are made sufficiently close to T_c. For example, if the constants M_0 are chosen so that the various curves have unit magnetization at a

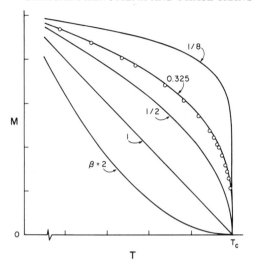

FIGURE 3.1   Power-law $M \propto |T_c - T|^\beta$, for various values of the exponent β. Data points are experimental measurements of the order parameter of the antiferromagnet MnF_2, in the range 1.8 degrees below the critical temperature $T_c = 67.336$ K.

temperature 10 K below T_c, then the curves for $\beta = 1/8$, 0.325, and 1/2 take on the respective values $M = 0.316$, 0.050, and 0.010, at a temperature 0.001 K below T_c. Thus there is a difference of a factor of 5 between the values in the last two cases.

The temperature variation of the order parameter on the coexistence curve is certainly not the only quantity that can be studied with experiments in critical phenomena. Another important quantity is the order-parameter susceptibility, defined as the derivative (i.e., the rate of change) of the order parameter with respect to a small change in the field to which it is coupled, while the temperature is held constant. For a magnetic system this quantity is the magnetic susceptibility (derivative of the magnetization with respect to magnetic field); for the gas-liquid critical point the order-parameter susceptibility is the iso-thermal compressibility (derivative of the density with respect to pressure, at constant temperature). These quantities become extremely large near the critical point, and we may write, for example, the zero-field magnetic susceptibility as

$$\chi = \chi_0/|T - T_c|^{\gamma}, \tag{2}$$

where the exponent γ is the same for all members of a universality class. The coefficient χ_0 varies from one system to another, and it is different above and below T_c; however, the ratio of its value above T_c to its value below T_c is a universal number—i.e., it is the same for all members of a universality class.

Another important quantity is the specific heat, defined as the derivative of the internal energy of the system with respect to a small change in temperature. The specific heat is found to become infinite at the critical point in some systems; for some other universality classes one finds that the specific heat is finite but has a sharp cusplike maximum at the critical point. In either case, one may define an exponent α that characterizes the anomalous behavior of the specific heat at the critical point. An example of the specific heat behavior is shown in Figure 3.2.

Although the critical exponents α, β, and γ defined above may be independent in principle, they were found empirically, in the 1960s, to obey a scaling law:

$$\alpha = 2 - \gamma - 2\beta. \tag{3}$$

This scaling law is one of the consequences of the more recent renormalization group theories.

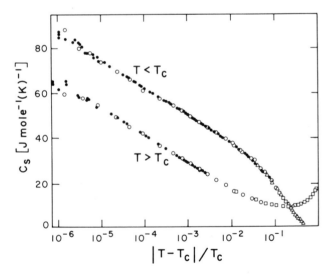

FIGURE 3.2   Specific heat C_s of liquid ^{4}He at saturated vapor pressure, near the temperature $T_c = 2.172$ K of the onset of superfluidity, as a function of $|T - T_c|$ on a logarithmic scale. A straight line on this plot would correspond to a logarithmic temperature dependence for C_s or a critical exponent $\alpha = 0$. Careful analysis of the data gives the result $\alpha = -0.026 \pm 0.004$.

A property of great interest near the critical point is the statistical correlation for fluctuations in the order-parameter density, at two nearby points in space, as a function of the distance between the points. This correlation function can be measured by neutron-scattering experiments in magnetic systems, and it can be measured by light-scattering experiments or small-angle x-ray-scattering experiments near the liquid-vapor transition. Near to the critical point, the correlation length, which characterizes the range of correlations for the order-parameter fluctuations, becomes extremely large relative to the typical spacing between atoms in the substance. This large correlation length is directly related to the large amount of long-wavelength fluctuations that were mentioned earlier and that give to critical phenomena their special subtleties and complexities. Naturally, there is great interest in studies of the variation of the correlation length with temperature, pressure, and other parameters, near the critical point.

It should also be mentioned that the integrated order-parameter correlation function, which can be directly measured by a scattering experiment in the limit of small angles, is related by a theorem of statistical mechanics to the order-parameter susceptibility χ defined above; thus a scattering experiment may be a convenient method of measuring χ. Also, for systems like an antiferromagnet, in which the order parameter describes a quantity that oscillates as a function of position in space, a scattering experiment may be the only direct way of measuring the value of the order parameter $M(T)$ in the broken-symmetry phase below T_c.

Many other experimental techniques have also been used to study properties of various systems near critical points. For example, the temperature coefficient of expansion of a solid, which can be measured with great precision, has similar behavior to the specific heat near a critical point; the index of refraction of a fluid has been used as a measure of its density; the rotation of polarized light by a transparent ferromagnet (Faraday effect) has been used to study the temperature dependence of the magnetization.

The quantities discussed above are all equilibrium or static quantities; they can be measured in a time-independent experiment, under conditions of thermal equilibrium, and any correlation functions involved refer to the correlations of fluctuations at a single instant of time. The majority of theoretical studies and of experiments on critical phenomena are concerned with these static measurements, and the usual division of systems into different universality classes is based on these static phenomena. There are other properties of systems, known as dynamic properties, which require a more detailed theoretical analysis and which require a further subdivision of the universality classes—i.e., two systems that belong to the same universality class for their static properties may show quite different behaviors in their dynamic properties. Examples of dynamic properties are various relaxation rates when the system is slightly disturbed from equilibrium, correlations involving fluctuations at two different instants of time, and transport coefficients, such as the thermal and electrical conductivities. Among the experiments used to study dynamic properties are measurements of sound-wave attenuation and dispersion, widths of nuclear or electron magnetic resonance lines, and inelastic-scattering experiments, in which the energy change of the scattered particle is determined along with the scattering angle.

Typically, one finds that the relaxation rate of the order parameter becomes anomalously slow at a critical point. Some other relaxation rates are found to speed up, however, and transport coefficients become large in a number of cases. In some cases, the results of a dynamic experiment may be interpreted as an indirect measurement of a static property of the system. In fact, some of the most precise measurements of static critical properties have been obtained by dynamic means. Examples here are measurements of the superfluid properties of liquid helium, the low-frequency sound velocity of a fluid, and the frequency of nuclear magnetic resonance in a magnetic system.

WHAT DETERMINES THE UNIVERSALITY CLASS?

We now know what determines the static and dynamic universality classes in most cases, although there remain a number of difficult cases that are not resolved. Since there is no simple rule that is completely general, we shall describe here only a few simple cases and give a few examples of factors that do or do not change the universality class of these systems.

Consider an idealized magnetic system in which magnetic atoms sit on the sites of an elementary periodic lattice—such as a simple cubic lattice. Each magnetic atom has an elementary magnetic moment or spin, of fixed magnitude, which can point *a priori* in one of several directions of space.

In the Ising model, which we have referred to several times above, we suppose that the atomic moments can point in only two directions— either parallel or opposite to some fixed axis, which we shall take to be the z axis. Then the microscopic state of this model is determined by specifying for each atom whether the magnetic moment along the z axis is positive or negative. In the Ising model we assume an interaction between nearest neighbor atoms, tending to align the moments in the same direction. At temperatures above the critical temperature, the aligning force cannot overcome the disordering tendency of random thermal motion, and there are equally many positive and negative moments. At temperatures below T_c, however, one of the two possible directions acquires a majority of the moments, and there is a net magnetization $\pm M(T)$ that measures the size and direction of this majority. Thus, the Ising model has a plus-minus symmetry that is broken below T_c.

Another model of magnetism is the Heisenberg model. Here it is assumed that the atomic moments can point in any direction of space. The aligning force between neighboring spins is assumed to be derived from an energy that depends only on the angle between the two spins and not on their orientation with respect to any fixed axis in space. The energy of this model is unchanged if we rotate all the spins in the system by the same angle, about any direction, and we say that the model is symmetric under arbitrary rotations of the spins. This symmetry is broken below T_c, when there is a magnetization that spontaneously picks out some direction in space.

A system intermediate between the Heisenberg model and the Ising model is the XY model, where the directions of the magnetic moments are restricted to lie in a single plane (say the XY plane). Again, the energy is taken to be unchanged if all spins are rotated by the same angle in this plane. The Ising model, the XY model, and the Heisenberg model may be said to have order parameters that are, respectively, a 1-D vector, a 2-D vector, and a 3-D vector.

We have already seen that spatial dimensionality is crucial in determining the universality class of a system—the Ising model on a 2-D lattice has different critical exponents from the 3-D Ising model, for example. The dimensionality (or symmetry) of the order parameter also turns out to be important. The critical exponents of the Ising model, XY model, and Heisenberg model in three dimensions differ from each other by a small but significant amount. (For example, the exponent β defined above takes on the values 0.325, 0.346, and 0.365 in the three cases.) In two dimensions the differences are more dramatic. We believe that the 2-D Heisenberg model has no phase transition at all—it remains paramagnetic (disordered) at all temperatures other than zero. The 2-D XY model is believed to have a phase transition of a peculiar type (see below), for which the critical exponents α, β, and γ cannot be properly defined.

We may next ask what happens if a model has a 3-D vector order parameter similar to the Heisenberg model, but the interactions give a lower energy to spins in the z direction than to spins in the other direction of space. In this case the order parameter has the symmetry of the Ising model, and the critical exponents are those of the Ising model rather than those of the Heisenberg model. If the energy favoring the z direction is small compared to the Heisenberg-type interactions favoring parallel alignment of spins without regard to the particular direction in space, then we expect to observe a crossover behavior: close to the critical point (say $|T - T_c| \leq 10^{-5} T_c$) we will see the critical exponents of the Ising model, but farther from T_c there may be a range of temperatures (say $10^{-2} T_c > |T - T_c| > 10^{-4} T_c$) where the system appears to have the critical exponents of the Heisenberg model. An experimentalist seeking to measure accurately the critical exponents of some universality class will naturally try to avoid using systems that have a crossover in the middle of the accessible temperature range.

In more complicated systems, with multicomponent order parameters, there is a variety of possible higher-order symmetry breaking terms, which may favor some discrete subset of the possible orientations of the order parameter. In some cases these terms lead to a change in critical behavior; in some others they lead to a small fluctuation-induced first-order transition, even though the classical theory predicts a continuous transition.

There are also many factors that are known to be irrelevant to deciding the universality class. The precise nature of the spatial lattice is unimportant—for example, an Ising model on a hexagonal lattice in two dimensions will have the same critical exponents as on a square or rectangular lattice. The exponents are also unaffected if the interactions are stronger along one spatial direction than another.

The universality class of a magnetic model is unchanged if the interaction between spins extends beyond nearest neighbors on the lattice, provided that the interaction falls off sufficiently rapidly with separation. In real magnetic systems, however, there is an important long-range interaction that does not fall off rapidly with distance—the magnetic dipole interaction, which decreases only as the inverse cube of the distance between atoms. This is sufficiently long range to change the universality class of a ferromagnet. Particularly in an Ising system, the dipole interaction has a drastic effect on the critical behavior. For a system like iron, where the magnetic dipole interaction is weak compared with the quantum-mechanical exchange interactions responsible for the ferromagnetism of the material, the dipole interaction only becomes important close to the critical point. However, there also exist cases where the dipole interaction is large and one readily sees an effect on the critical behavior. Neutron scattering and specific-heat measurements on one of these systems ($LiTbF_4$) have provided dramatic confirmation of the peculiar critical behavior predicted theoretically for the Ising model with dipolar forces.

We may remark here that the long-range magnetic dipole interactions are irrelevant to the critical behavior of antiferromagnets, because of the cancellations arising from the alternating directions of the spins in this case.

We have already noted that the liquid-vapor critical point has the same critical exponents as the 3-D Ising model. The liquid-vapor order parameter, which we take as the difference from the density at the critical point, is a real quantity, which can be positive or negative like the magnetization of the Ising model, but the fluid does not possess a precise symmetry between positive and negative values of the order parameter. It is a prediction of renormalization group calculations that this remaining asymmetry is irrelevant for the critical behavior, and indeed experiments confirm with high precision the identity of the critical exponents for the fluid and Ising critical points.

In two-component fluid mixtures, there is often a critical point for phase separation, which is closely analogous to the liquid-vapor critical point. This critical point also falls in the Ising universality class, and it has been studied in many experiments.

The critical behavior of the XY model is particularly interesting because this model is predicted to fall in the same universality class as the superfluid transition of liquid helium (^{4}He). In the latter case the order parameter is a complex number representing the quantum-mechanical condensate wave function of the superfluid, and the relation to the XY model results from the mathematical representation of a complex number as a vector in the XY plane. Because liquid ^{4}He can be obtained with great purity, and because temperatures near the superfluid transition can be controlled with high precision, critical exponents have been measured with high accuracy in this system. The excellent agreement with calculations for the XY model provide both a confirmation of the modern theory of critical phenomena and an important confirmation of the theory of superfluidity as well.

EXPERIMENTAL REALIZATIONS OF LOW-DIMENSIONAL SYSTEMS

Although the world we live in is three dimensional, theoretical studies of 2-D systems have direct applications to systems in nature. For example, a transition between commensurate phases of a layer of atoms adsorbed on a crystalline substrate, or the melting of a commensurate adsorbate phase, will generally fall into the same universality class as some simple 2-D model with a discrete order parameter, such as the 2-D Ising model or the three-state Potts model (an Ising-like model, realized by some gases adsorbed on graphoil, in which each spin can take on three, rather than two, values; the energy of an interacting pair of spins is lower if they have the same value, and higher if they are different). Recent experimental developments, including improved substrates, and the availability of synchrotron x-ray sources have made possible new precise measurements of phase transitions in adsorbed gas systems.

In general, a film that is extended in two dimensions, but thin in the third dimension, will show the same critical behavior as some 2-D models. Slightly thicker films may show a crossover behavior from 3-D behavior to 2-D behavior as one gets closer to the critical point. Two-dimensional behavior can also be studied in 3-D layered systems, when the interactions between layers are sufficiently weak. In this case, one typically sees a crossover from 2-D to 3-D behavior as one gets closer to the critical point.

Interesting phenomena also occur in quasi-1-D materials such as crystals with chains of magnetic atoms and only weak interactions between chains, even though a true 1-D system does not show a phase transition at finite temperature. There has been a variety of experiments in quasi-1-D and quasi-2-D systems that demonstrates the expected crossover behaviors.

MULTICRITICAL POINTS

Although the gas-liquid critical point of a pure fluid occurs at a single point in the pressure-temperature plane, this same critical point becomes a critical line in the three-parameter space of pressure, temperature, and composition in the case of a two-component mixture. The transition temperature of an antiferromagnet may also become a line of critical points when a uniform applied magnetic field is included as a parameter.

There exists in nature a variety of special multicritical points, where several lines of critical points come together. Multicritical points have been studied experimentally in multicomponent fluid mixtures, in magnetic systems, and at the tricritical point of superfluidity and phase separation in a liquid ^{3}He-^{4}He mixture.

SYSTEMS WITH ALMOST-BROKEN SYMMETRY

Some of the most interesting phase transitions involve systems in which the low-temperature phase has a special type of order, where there is almost, but not quite, a broken symmetry. It had been noted, as early as 1930, that thermally excited long-wavelength fluctuations should have the effect of destroying the long-range order and the broken symmetry of certain types of 2-D systems. (These include the 2-D Heisenberg and XY magnets and the 2-D superfluid.) It was noted similarly that thermally excited long-wavelength vibrational modes must destroy the periodic translational order of a 2-D crystal. Although, for many years, it was believed that the absence of broken symmetry implied, in turn, the absence of a phase transition in all those cases, this conclusion began to be questioned in the 1960s. In particular, it was proposed that the XY magnet, the superfluid, and the crystal might have a distinct low-temperature state, in two dimensions, where the order parameter has a kind of quasi-order, in which there are correlations over arbitrarily large distances that fall off only as a small fractional power of the separation between two points. This behavior has recently been proven with complete mathematical rigor, in the case of the XY model at low temperatures. Since this power-law behavior is different from the exponential falloff of the correlation function (short-range correlations) that one finds at high temperatures in the same systems, there must be a definite temperature separating these two behaviors, which is by definition a phase transition temperature. (In the 2-D Heisenberg model, however, it is believed that there are only short-range correlations at all temperatures above zero, and, hence, there is no phase transition.)

Two-Dimensional Superfluid and XY Model

In the 1970s, there was developed a theory in which the transition to short-range order in the 2-D XY model and superfluid occurs as a continuous transition (i.e., not first order) that can be described by the proliferation of point-like topological defects in the order parameter of the system. This theory makes a number of specific predictions about both static and dynamic properties near the phase transition, which differ significantly from the behaviors at other critical points. These predictions have been confirmed to some extent by experiments on thin films of superconductors and of superfluid helium and by numerical simulations of the 2-D XY model, but the accuracy of these comparisons is not yet sufficient to be considered incontrovertible support for the theory.

Melting of a Two-Dimensional Crystal

An application of the point-defect mechanism to the melting of a 2-D crystal again makes a number of striking predictions, the most interesting of which is that there should be a new hexatic liquid-crystal phase, existing in a narrow temperature region between the crystal and the isotropic liquid. The hexatic phase would possess quasi-order in the orientation of bonds between neighboring atoms but would have only short-range correlations in the positions of atoms, as one finds in the true liquid state. Many workers in the field believe, however, that some other melting mechanism (perhaps grain boundaries) will necessarily intervene and produce a first-order melting before the melting temperature for the point defect (dislocation) mechanism is reached. A first-order transition could make it impossible to reach the hexatic liquid crystal phase. (In 3-D bulk systems, melting is always a first-order transition.)

This question continues to generate controversy, as computer simulations tend to favor a first-order transition, while scattering experiments on incommensurate crystalline layers of argon, xenon, or methane, adsorbed on a graphite substrate, provide strong evidence for a continuous melting transition, for some range of coverages. The xenon experiments also provide evidence for the existence of a hexatic phase. Further work is necessary, however, particularly to clarify the possible effects of the crystal substrate on the melting transition.

In layered phases of certain organic molecules (smectic liquid crystals), there are phase transitions arising from a change in the order within a layer, which may be considered as generalizations of the 2-D melting transition. A phase describable as a stack of hexatic layers has been observed by x-ray experiments in several smectics.

Smectic A-to-Nematic Transition

Although the most elementary examples of systems with almost-broken symmetry (quasi-order) are the 2-D systems discussed above, the phenomenon also occurs in certain 3-D liquid-crystal phases. The simplest of these is the smectic A phase, in which long organic molecules are arranged with their axes parallel to a particular direction of space, and, in addition, the centers of gravities of the molecules tend to be arranged in a series of equally spaced layers perpendicular to the molecular axes, i.e., there is a periodic modulation of the density in one direction. In the directions parallel to the planes there is only short-range, liquidlike order.

Because it costs little energy to excite long-wavelength bends in the molecular layers, thermal fluctuations in these modes are large, and the resulting displacements reduce the periodic translational order to quasi-long-range order (power-law correlations), as in a 2-D solid.

When heated, a smectic A may lose its remaining translational order, while retaining the orientational order of the molecular axes. We then say the material has undergone a transition to the nematic phase. The theory of this phase transition is complicated by the coupling between the translational- and orientational-order parameters and also by the large differences in the microscopic properties of the nematic phase, when measured along different directions.

Experimentally, well-defined critical exponents have been seen, holding over several decades in the distance from the transition temperature, for such properties as the translational correlation length, measured by x-ray scattering, in the nematic phase. However, the critical exponents are different in the directions parallel and perpendicular to the molecular axis and also vary from one material to another.

QUENCHED DISORDER

The discussion, until this point, has focused on systems in thermal equilibrium, where the important fluctuations arise from the intrinsic thermal population of excitations, required by the laws of statistical mechanics. In many solid-state systems of interest, however, there may be additional, frozen-in disorder quenched into the system on formation of the sample. Depending on the nature of the phase transition, the nature of the quenched disorder, and the way in which the quenched disorder couples to the order parameter of the phase transition, the critical behavior may or may not be changed by the disorder. In the most extreme cases, the phase transition may be smeared out or eliminated entirely. The effects of quenched disorder appear to be well understood in many cases, but there remain others that are poorly understood and are the subject of active investigation.

One particularly interesting case occurs when the quenched disorder couples linearly to an Ising-like order parameter, as would be the case if there were a local magnetic field of random sign on each site of the Ising ferromagnet. (This situation has been realized experimentally in an Ising-like antiferromagnet, with a uniform magnetic field and randomly missing magnetic atoms.) Different theoretical approaches have led to opposite expectations for whether or not there should be a sharp phase transition in this case, and experimental measurements have not yet resolved the issue in a satisfactory manner. The question may also have implications for elementary-particle physics, because one of the theoretical approaches takes advantages of a close mathematical analogy between the random magnetic-field problem and so-called supersymmetric models in quantum field theory.

Extreme examples of systems with quenched disorder are the spin-glass phases, discussed in Chapter 4. There are many open questions related to phase transitions in these systems.

PERCOLATION AND THE METAL-INSULATOR TRANSITION IN DISORDERED SYSTEMS

There are a number of problems in condensed-matter physics that bear a qualitative resemblance to systems at a continuous phase transition and that may indeed be understood by methods of analysis similar to those used in the theory of critical phenomena but where the source of disorder is entirely quenched randomness and not thermal fluctuations. Among these are various problems concerned with geometry and transport in disordered systems, including metal-insulator transitions in disordered systems where quantum mechanics plays a critical role, as well as the classical problem of percolation in a mixture of macroscopic conducting and insulating particles.

NONEQUILIBRIUM SYSTEMS

A number of problems resembling critical phenomena have been observed in systems out of equilibrium. As one example, there have been experimental and theoretical studies of the phase-separation critical point of a binary fluid mixture, under strong shear flow.

The renormalization group concept has already had a profound impact on the theoretical ideas and mathematical techniques (e.g., iterative maps) used to describe changes of the state of motion in nonequilibrium fluids, at moderate Reynolds numbers (Chapter 11). Turbulence in fluids at high Reynolds numbers has certain features of universality and scaling that suggest that theoretical techniques used to understand critical phenomena may also have a bearing here.

FIRST-ORDER TRANSITIONS

In general, one does not find at a first-order phase transition the rich variety of phenomena that one finds at a critical point, and first-order transitions have consequently received relatively less attention in recent years. Several classes of universal phenomena associated with first-order transitions do deserve mention, however, because they have been the subject of continuing research, and because they still contain outstanding puzzles. These include the areas of nucleation phenomena, limits of superheating and supercooling, spinodal decomposition, and mathematical questions concerning the nature of the singularities in thermodynamic functions at a first-order transition.

The past few years have seen significant advances in research on the dynamics of phase transitions in fluid mixtures, but many basic questions remain unanswered. It is now clear that the transition from spinodal decomposition (from an unstable region) to nucleation and growth (from a metastable region) is a gradual one; there is no abrupt changeover at a spinodal curve. Some progress has been made in providing a general theoretical description of the phase separation process, and machine and laboratory experiments have provided some clues about the underlying physics. A global scaling procedure, first recognized in computer simulations, has been described by simple models whose validity has been demonstrated experimentally. A long-standing doubt about the validity of nucleation theory in the neighborhood of critical points has apparently been laid to rest by a theory that demonstrates that the anomalous behavior is the result of critical slowing down of growth and by experiments that are consistent with the theory.

Of course, there are many important open questions of a nonuniversal nature concerning first-order transitions, as there are for continuous transitions. These include such matters as understanding the microscopic mechanisms for various transitions and calculations of the location of the transition and of the sizes of the discontinuities of various physical quantities. Such questions are discussed elsewhere in this report, in the chapter appropriate to the particular transition. Indeed, the reader will find that phase transitions are featured in virtually every chapter of the report.

OUTLOOK

Work on critical phenomena and related problems, in the 1980s, should lead to progress in a number of directions, among which we may expect the following:

1. There will be more precise experimental tests of the predictions of the renormalization group theories and of some of its consequences (e.g., scaling laws among exponents and the universality of certain relations among absolute values of properties near T_c). This is necessary, because it is important to test thoroughly the underpinnings of a theoretical approach that is seeing such widespread application in modern physics.

2. There will be an extension of our understanding to some of the

more complicated types of critical points. We may also expect a better understanding of crossover phenomena and, more generally, of the corrections to simple power-law behavior that are necessary to understand measurements that are some distance away from the critical point. Combinations of renormalization-group approximations and accurate microscopic models will be used increasingly to calculate entire phase diagrams, including the locations of first-order transitions, in a variety of systems.

3. Some of the outstanding problems mentioned above may be solved, perhaps by means of some important new calculational methods, or perhaps by the development of some new physical ideas, or by refinements in experimental techniques. For example, it seems likely that in the 1980s considerable progress will be made in our understanding of 2-D melting and related phenomena, through experiments on a variety of systems, including liquid crystals, both in bulk and in suspended films of several layers thickness; adsorbed layers; the electron crystal on the surface of liquid helium; and perhaps synthetic systems, such as a film containing colloidal polystyrene spheres. The role of the substrate in the transition, in the case of adsorbed layers, will be investigated. The construction of a theory of the smectic A-to-nematic transition is one of the most challenging unsolved problems in critical phenomena. Interest in this problem is heightened by its possible connection to phase transitions in idealized superconductors and in certain quantum-mechanical models of interest to elementary-particle theories. A variety of other phase transitions among other liquid-crystal phases is also poorly understood at present and will undoubtedly be the subject of major investigations in the next few years. Problems of disordered systems in which the disorder is quenched into the system during its formation, and is not due to thermal fluctuations, remain an important area where new ideas are necessary, and the theoretical methods of critical phenomena need further development.

4. The theoretical and experimental methods used to study the problems of critical phenomena will be applied to the study of other problems in condensed-matter physics.

4

Magnetism

INTRODUCTION

With respect to their magnetic properties, solids can be divided into two categories depending on the direction of the moment induced by an applied magnetic field. If the moment is opposite in direction to the field, the material is said to be diamagnetic; materials where the moment is parallel to the field are paramagnetic. Apart from its role in superconductivity, diamagnetism is a weak effect. In contrast, concentrated paramagnetic materials often display very large responses corresponding to local fields of the order of several hundred tesla. At present most of the research in magnetism involves systems that show paramagnetic behavior.

In paramagnetic materials the magnetism is associated with partially filled inner shells of atomic electrons. These are the $3d$ shell (transition metal compounds), the $4f$ shell (rare-earth compounds), and the $5f$ shell (actinide compounds). The classification can be extended further depending on whether the electrons in the partially filled shells are localized, as happens in insulators and semiconductors, or itinerant, as in many metals. In magnetic insulators each atom with an unfilled shell possesses an intrinsic magnetic dipole moment. In addition to their interaction with the applied field, the dipoles interact with one another through long-range dipolar forces and short-range exchange interactions, the latter arising from the interplay of electrostatics and quantum

95

symmetry. Besides their mutual interactions, the moments also inter-act with the charges on the surrounding ions, which create a crystalline electric field. The moments in itinerant magnets are delocalized, extending throughout the system. As in the case of localized moments, there are both dipolar and exchange interactions.

Strictly speaking, in many materials the paramagnetic behavior alluded to earlier is observed only at high temperatures. As the temperature is lowered the system undergoes a phase transition to a state characterized by long-range magnetic order. The most familiar example of this behavior occurs in ferromagnets where the long-range order appears as a spontaneous magnetization. The transitions are caused by the exchange interactions between the moments and gener-ally occur at temperatures comparable to the strength of the interaction between neighboring moments.

Studies of the magnetic properties of materials with localized mo-ments occupy an unusual position in solid-state physics. This happens because these properties can be characterized using models involving only the individual atomic moments, rather than the full array of atomic electrons. These models, generally referred to as spin Hamiltonians, have generic forms that depend on the symmetry of the system and interaction parameters that reflect the microscopic environment of the magnetic ions. The study of spin Hamiltonians is a central part of many-body theory and statistical mechanics. The spin Hamiltonian formalism makes possible a direct connection between real materials and formal theory that has been a driving force for much of the research in magnetism in recent years. Experimental studies have provided crucial tests of theories of collective excitations and critical phenomena. In turn, the availability of accurate data has stimulated the development of increasingly precise theories.

The past decade has been one of remarkable progress in magnetism. By 1970 it can be said that the behavior of isolated magnetic ions in insulators was well understood. The low-temperature properties of ideal, three-dimensional (3-D) magnetic insulators were successfully interpreted in terms of elementary excitations. The high-temperature behavior had also been investigated, and theoretical studies utilizing high-temperature expansions had given useful insights. There was growing interest in magnetic critical phenomena, but no unifying microscopic theory was available. The magnetic properties of metallic systems were not well understood. The behavior of isolated magnetic ions in nonmagnetic metallic hosts had revealed unexpected complex-ities at low temperatures. In addition, insight into the behavior of common magnets like iron and nickel had not advanced much beyond the level of molecular field theory. As will be discussed below, in the intervening years there have been significant advances in our understanding of both insulating and metallic magnets. These occurred not only in ideal systems but in various types of disordered materials as well.

The period since 1970 has also been an era of rapid growth in computer simulation studies of various static and dynamic properties of magnets. Such a development would have been impossible without large, high-speed digital computers.

In addition to its place in basic research, magnetism continues to make important contributions to technology through improved and novel materials for information storage and power generation, for example.

MAGNETIC INSULATORS

Low-Dimensional Systems

Recently there has been a shift in emphasis away from studies of ideal, 3-D magnetic insulators. Increasing attention has been directed toward low-dimensional (low-D) systems, an area of research that has grown rapidly in the past 15 years. Before this period, the most important single research achievement in the field was undoubtedly the famous Onsager solution of the two-dimensional (2-D) Ising model. The Ising model is a simplified version of a physically realistic model of magnetism. The importance of the Onsager solution is that it shows that a phase transition is possible in a simple model with short-range magnetic interactions, a matter previously open to doubt. Furthermore, the nature of the critical behavior is significantly different from the older, molecular-field type of approximate theories of critical behavior in magnets. In the 40 years since its appearance in 1944, the Onsager solution has been exploited in a variety of ingenious ways, which have provided much of the basis of the modern theory of critical phenomena.

The one-dimensional (1-D) version of the Ising model was solved back in 1925 but was not considered interesting because the solution did not show a phase transition. Around the early 1960s, however, a number of other 1-D models of magnetism were solved, either exactly or numerically. At that time, the feeling was widespread that 1-D models were merely amusing toys for mathematical physicists to play with, with little or no relation to the real, 3-D, physical world. Nevertheless, the appearance on the scene of a number of 1-D model solutions attracted the attention of experimental physicists, who searched successfully for chemical systems with highly spatially anisotropic magnetic properties. They were able to show that the experimental behavior of the real systems agreed well with that of the 1-D theoretical models. Subsequently, the process of molecular engineering was developed, whereby quasi-2-D and quasi-1-D magnetic systems were prepared according to specifications by inserting large nonmagnetic spacer molecular complexes (usually organic) into suitable systems to increase the physical separation between planes or chains of magnetic ions, respectively. In this way, the magnetic interactions were substantially reduced in one or more crystalline directions.

As noted, low-D physics is continuing to grow in importance relative to the traditional 3-D variety for the following reasons:

1. In general, the ease of solution of a particular model of cooperative magnetism increases as the dimensionality decreases. Hence, a variety of exact solutions of varied and nontrivial models is now available in 1-D, whereas there are hardly any exact solutions in 3-D. (The value of exact solutions can hardly be overestimated.) A useful secondary feature of 1-D exact solutions is their ability to serve as testing grounds to give insight into the degree of reliability of the various approximate calculational techniques that must, of necessity, be employed in 3-D.

2. A feature of great importance is the wealth of novel and interesting physical phenomena that are peculiar to low-D. Examples include prominent quantum effects in low-D, e.g., in the area of low-temperature spin dynamics, and the enhancement, by virtue of topological considerations, of the effects of impurities and randomness in low-D. Further, the current trend in physics is to move away from the traditional approach of the linear (harmonic) approximation to consider nonlinear effects. In the linear approximation to a model magnetic system, the small-amplitude collective excitations are called magnons, and their behavior has been studied for decades. Recently, the importance and interesting properties of nonlinear (large-amplitude) excitations have been recognized. Various types of such phenomena exist, called, for example, solitons, kinks, vortices, breathers, instantons, and domain walls. These excitations are important in many areas of physics, including plasma physics, turbulence, and field theory, but they appear to be most easily investigated, theoretically and experimentally, in magnetic systems, particularly in 1-D, but also in 2-D, systems (Figure 4.1).

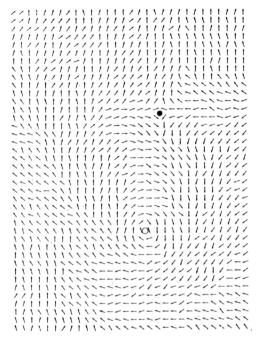

FIGURE 4.1 Nonlinear vortex excitations in the two-dimensional *XY* model. The dark and open circles denote the centers of spin vortices of opposite circulation. [S. Miyashita, H. Nishimori, A. Kuroda, and M. Suzuki. Prog. Theor. Phys. *60*. 1669 (1978).]

3. A fascinating new development of recent years is a phenomenon that may be termed, for convenience, mapping. Mapping refers to the discovery that apparently different physical phenomena are, in fact, related to one another through an underlying mathematical description. The same mathematics has been found to describe a variety of physical systems, with appropriate definitions of the relevant mathematical parameters. This result may be characterized as obtaining several solutions for the price of one. Mappings have been discovered between systems of the same or different dimensionalities. A well-known example of the former case is the class of systems that are isomorphous to the 2-D *XY* model of magnetism. This class includes 2-D superfluids, 2-D melting solids, smectic (layered) liquid crystals, and 2-D Coulomb gases.

Possibly the most famous mapping between systems of different dimensionality involves the model many-body system consisting of a single magnetic impurity exchange-coupled to a sea of conduction electrons. Experimentally, dilute solutions of such impurities in otherwise normal metals display noticeable anomalies in susceptibility, in specific heat, and in their temperature-dependent resistivities. These anomalies are referred to collectively as the Kondo effect. Their explanation constitutes the Kondo problem. In a remarkable development, the 3-D Kondo problem has been mapped onto a solvable 1-D quantum model. The calculated susceptibility and specific heat agree well with experiment. At the same time, it is a tribute to the power of the renormalization group method (Chapter 3) that the solution of the Kondo problem obtained through its use, though manifestly an approximation, is nevertheless demonstrably accurate when compared with the exact solution.

The Kondo mapping is presumably the first of many mappings from 3-D to a much more tractable lower dimensionality. This factor, together with the fact that new mappings are turning up in rapid succession, makes low-D physics applicable to more areas than one might, at first sight, suppose.

Critical Phenomena

As noted in Chapter 3, the 1970s was a period of intense activity in the field of phase transitions and critical phenomena. Studies of phase transitions in magnetic materials, primarily insulators, confirmed many of the predictions of high-temperature series and renormalization group calculations. In addition, they provided important evidence in support of the concepts of scaling and universality. In the first part of the decade most of the research pertained to ideal, 3-D magnets. Currently, greater emphasis is being placed on studies of critical phenomena in disordered and lower-D systems.

METALLIC MAGNETS

Transition-Metal Ferromagnets

Metallic magnets can be divided into two classes depending on whether the magnetic atoms belong to the transition-metal series or to the rare-earth and actinide series. Recent advances in the theory of transition-metal ferromagnets have led to a better understanding of the nature of their ground state and of their magnetic properties at finite temperatures.

1. After 50 years of discussion, it is now widely accepted that the ground state of iron, nickel, and most other transition-metal fer-

romagnets is best described by an itinerant or band picture, as opposed to a localized picture. de Haas-van Alphen measurements have generally agreed with the results of band calculations. The calculations themselves have now been greatly improved by using better algorithms and larger machines. The greatest advance, however, is the use of a new theory that incorporates electron-electron effects into the single-particle states. Such calculations now correctly predict which transition elements are ferromagnetic, and they give improved agreement with Fermi surface data.

These band calculations are also used to determine cohesive energies and bulk moduli of whole series of materials with great success. In particular, the anomalously large lattice constants of the magnetic transition metals, as compared with the trend of their nonmagnetic neighbors, can be understood from the computed magnetic components of their cohesive energies. Another extension is to the calculation of the spin-wave scattering, which also agrees well with measurements.

A stringent test of the band picture is given by angle-resolved photoelectron spectroscopy, which determines both the energy and wave vector of the emitted electron. An improvement to include spin-polarization analysis was recently announced (Figure 4.2). Large-scale angle-resolved measurements became feasible with the advent of high-luminosity synchrotron radiation sources (although some important high-resolution work is done with conventional sources). By and large, the measured dispersion curves are in agreement with band calculations, giving direct evidence for band states in a wide class of materials, magnets in particular. There is, however, still much to be done to achieve an understanding of the various effects of the real holes created in the photoemission process. A beginning has been made, but the problems are formidable.

2. The natural model for the temperature dependence of band ferromagnetism is that of Stoner, which is unsatisfactory, at least for some materials, in several respects. It predicts too high a Curie temperature; it does not incorporate directly the thermodynamically dominant spin-wave excitations; and it does not admit the persistence of magnetic correlation effects above the Curie temperature. That such effects exist is shown by the Curie-Weiss susceptibility in the paramagnetic phase and, more dramatically, by the continued existence of spin-wave excitations far above T_c in iron and nickel that are seen in inelastic neutron scattering.

These observations and others have led to a number of new theoretical schemes for extending the ground-state band picture to finite temperatures. The competition between these schemes has revived the localized versus itinerant controversy in a new form.

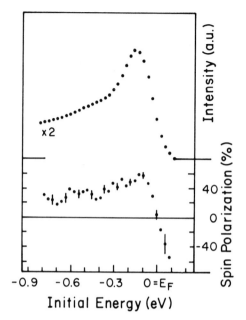

FIGURE 4.2   Electron distribution curves and corresponding spin polarization for photoemission from the (110) face of nickel. [R. Rune, H. Hopster, and R. Clauberg, *Phys. Rev. Lett. 50*, 1623 (1983).]

Paramagnon, or weak itinerant, models consider magnetic fluctuations about a basically nonmagnetic state. Though useful for enhanced paramagnets, and quite possibly for weak ferromagnets, they provide less correlation than is needed to describe iron and nickel. Phenomenological attempts to extend the paramagnon models to strongly correlated cases have been made and have met with some success.

Two approaches more directly concerned with the underlying electronic structure are the local-band theory and the alloy analogy. Each assumes a disordered magnetization configuration, approximately solves for the electronic states in the mean exchange field produced by the configuration, and demands that the configuration be reproduced self-consistently. The local-band scheme assumes that the important configurations are characterized by a short-range order, sufficient to define the exchange split bands locally even above T_c. In the alloy analogy the magnetizations at different sites are statistically independent, and electronic states are computed in the coherent potential approximation. Evidence in favor of short-range order at high temperature has been reported, but the interpretation of the results has been challenged. Currently, then, the contention is between the picture that the magnetization (although not the electrons making it up) is localized for a relatively long time at a single site, and the picture that it has coherent structure on a larger, 10-15 Å scale.

Rare-Earth and Actinide Magnets

There are three great conceptual distinctions between f (rare-earth and actinide) magnetism and d (transition-metal) magnetism. One is that, overall, orbital (as opposed to spin) magnetic effects are qualitatively more apparent in f magnetism. This shows up in many properties where the magnetic behavior has peculiarities associated with the coupling of the orbital moment to the crystalline lattice. Second is that in metallic systems the f electrons tend to delocalize as one moves toward the light end of the $4f$ or $5f$ row. Thus at cerium in the rare earths, or at plutonium, neptunium, and uranium in the actinides, one can study and hope to understand the effects involved in the transition from localized (Gd-like) to itinerant (Ni-like) magnetism. The lattice property most obviously correlated with this transition is the lattice parameter (or more strictly the f atom to f-atom spacing); however, the effects of the detailed electronic structure can significantly alter this correlation. Third is that the proximity of a sharp $4f$ level to the Fermi energy can lead to instabilities of the charge configurations (valence) and the magnetic moment.

There have been striking conceptual advances in the past decade associated with all three of these distinctive features of f-electron behavior. A discussion of these advances and their interrelationships follows.

1. The most characteristic consequence of the orbital contribution to the moment is strongly anisotropic magnetization behavior, with related peculiarities of magnetic structures and excitations. In the past decade this has been strikingly evidenced in cerium metallic and semimetallic compounds and, more recently, in the actinides, with most recent work in plutonium compounds. The association of exceptionally strong anisotropy in magnetic properties with the region where the local-to-nonlocal f transition occurs suggests a strong connection between the two phenomena. The availability of single crystals of actinide compounds, including those containing plutonium, has been a key element in making possible these advances.

2. Much of the intellectual excitement in f-electron magnetism in the past decade has been associated with a shift in experimental emphasis

from heavier rare-earth systems to cerium systems and into the light actinides. This excitement has arisen out of a variety of striking experimental observations, which in one way or another have tended to relate to the central question of the f-electron localized-to-delocalized transition. A variety of experimental techniques has been important in this conceptual opening. They have included high field magnetism, elastic and inelastic neutron scattering, and electromagnetic (e.g., de Haas-van Alphen, optical) and electron emission spectroscopies. On the theoretical side there have been two major advances. One of these has emerged from electronic (band) structure studies of the actinide elements, showing a transition in f-electron behavior from nonbonding (localized) at americium to bonding (delocalized) at plutonium. The other major theoretical advance has come out of Anderson-lattice model and band calculations pertinent to cerium and light actinide metallic, semimetallic, or semiconducting compounds. This theory shows that when the f electrons are moderately delocalized (intermediate between localized and band behavior, so as to be slightly bonding), f-electron-band electron hybridization (mixing) can explain a variety of otherwise anomalous properties including extreme anisotropy of magnetization and highly unusual magnetic structures and transitions.

3. The past decade has seen the birth and coming to fruition of a major area of research in valence instability. This interest was initiated by high-pressure experiments on samarium compounds at the beginning of the 1970s. Resistivity, volume change, and susceptibility measurements, done in rapid sequence, indicated the occurrence of a valence change of the samarium, with consequent semiconductor-to-metal and magnetic-to-nonmagnetic transitions. This, in turn, was followed by a vigorous research effort that showed that the candidate rare earths for mixed-valence behavior are Sm and Eu at the middle of the $4f$ row, Tm and Yb at the end of the row, and Ce at the beginning. The necessary condition for mixed-valence behavior is that two bonding states (i.e., with different f-occupation numbers) of the rare earth in the solid be nearly degenerate (Figure 4.3). It has become clear that the cause of mixed-valence behavior in cerium materials is different from that in the heavier rare earths. For the heavier rare earths mixed valence involves a fluctuation between two degenerate nearly localized states, whereas for cerium the mixed-valence behavior may be associated with a $4f$ localization/delocalization transition. Whether this is indeed so is a question of great interest. In the coming period, we can expect to see a lively search for mixed-valence behavior in actinide systems. It will be important to see whether light actinide

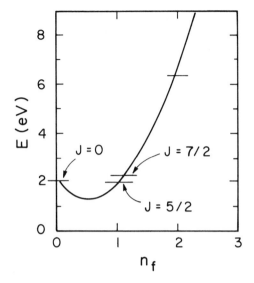

FIGURE 4.3   Schematic energy-level diagram for the intermediate valence Ce atom. n_f is the number of f electrons and J denotes the total angular momentum. [D. M. Newns, A. C. Hewson, J. W. Rasul, and N. Read, J. Appl. Phys. *53*, 7877 (1982).]

systems characteristically show cerium-type or samarium-type mixed valency, or a mixture of the two. On the theoretical side, we anticipate an exceptionally active effort in trying to understand the ground-state properties of lattices of mixed-valence ions on the basis of Hamiltonians that include narrow f states, broad conduction bands, hybridization, and f-f correlation effects.

DISORDERED SYSTEMS

Introduction

As in other fields of condensed-matter physics, the study of disordered materials has been an area of intense activity in magnetism in recent years. In ideal magnets the atoms are arranged on a lattice. The lattice structure is characterized by translational invariance. This is to say, atoms that are separated by one or more fundamental repeat distances have the same local environment. The existence of this invariance often simplifies the theoretical analysis, especially at low temperatures, where there is negligible thermal disorder.

In disordered magnets the translational invariance can be broken in different ways. The underlying lattice structure of substitutionally disordered materials is preserved. However, the sites of the magnetic atoms are occupied at random either by one of two (or more) different species of magnetic atoms, as in $Fe_xMn_{1-x}F_2$, or by either a magnetic or nonmagnetic atom, as in $Cd_{1-x}Mn_xTe$. In amorphous magnets the lattice is absent altogether. In this case the material is said to be topologically disordered. Examples of materials that can be prepared in the amorphous state include YFe_2, $(Fe-Ni)_{80}P_{14}B_6$, and $Fe_{1-x}B_x$. Usually the preparation involves rapid quenching from the melt so as to avoid crystallization.

The fundamental problem in the study of disordered magnets is to understand the effects of the disorder on the magnetic properties. If the disorder is weak, i.e., if only a few nonmagnetic atoms are present in an otherwise fully occupied magnetic lattice or a low concentration of magnetic atoms is present in a nonmagnetic host, one can interpret the behavior as a superposition of effects due to isolated impurities. Although the study of impurities is an important topic in its own right, the main emphasis currently is on highly disordered systems where an analysis based on the single-impurity picture is not applicable.

At high temperatures thermal disorder generally dominates any substitutional or topological disorder with the consequence that ideal and disordered magnetic materials behave in a qualitatively similar manner. However, as the temperature is lowered toward the regime where the energy associated with the thermal fluctuations becomes comparable to the strength of the interactions between the individual moments, the absence of translational invariance becomes increasingly important. The question then arises as to whether there is a transition out of the high-temperature phase. Should this be the case, is it to a state of conventional magnetic order or to a low-temperature disordered phase not present in the ideal magnets?

Disordered Ferromagnets, Antiferromagnets, and Paramagnets

The title of this subsection refers to systems that undergo phase transitions to states of conventional long-range order characteristic of ideal magnets or else are sufficiently dilute that they remain in their high-temperature or paramagnetic phase at all temperatures. An important question pertaining to those systems that do undergo transitions is the influence of disorder on the critical temperature, critical indices (Chapter 3), and other properties characteristic of the transition.

The behavior of the disordered systems at low temperatures is also interesting. As with the ideal magnets one can interpret the static and dynamic properties in terms of a nearly ideal gas of magnon excitations. Even in the lowest-order, or linear, approximation the calculation of the spectrum of excitations is a formidable problem. Nevertheless considerable progress was made toward its solution in the past decade. This progress came about in a number of ways. Experimental studies involving inelastic neutron and light scattering have provided detailed information about the magnons in various disordered magnets. Paralleling the experimental work there have been two types of theoretical investigation. The first involved purely analytic work mostly under the general heading of the coherent potential approximation, a name that reflects its origin as an approximation introduced in the calculation of the electronic properties of disordered alloys. The second entailed the development of computer simulation techniques, which made possible a direct calculation of the neutron-scattering cross section by integrating the linearized equations of motion of the spins (Figure 4.4).

Studies of magnons in substitutionally disordered magnetic insulators have provided important general tests of our understanding of

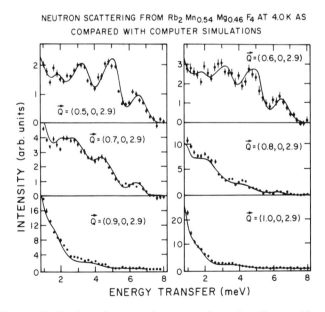

FIGURE 4.4 Distribution of scattered neutrons from the dilute antiferromagnet $Rb_2Mn_{0.54}Mg_{0.46}F_4$. The solid lines are computer simulations. [R. A. Cowley, G. Shirane, R. J. Birgeneau, and H. J. Guggenheim, Phys. Rev. B *15*, 4292 (1977).]

collective excitations in disordered systems. The reason for this is that the interactions in insulators are of short range so that the model spin Hamiltonian is characterized by only one or two parameters, which can often be inferred from independent measurements. In such a situation differences between experiment and theory cannot be explained away by a suitable adjustment of the parameters. This situation contrasts with studies of the electronic states and lattice vibrations in disordered materials. These provide a much less stringent test of theories like the coherent potential approximation since there are many more unknown, and hence potentially adjustable, parameters in the models.

One of the most interesting topics in the area of dilute magnetic insulators concerns their behavior near the critical percolation concentration. The percolation concentration refers to the concentration below which there is no longer an infinite cluster of mutually interacting magnetic atoms. Near the percolation point there is a direct competition between the thermal disorder due to the temperature and the substitutional disorder coming from the dilution. In addition to the critical behavior one is also interested in the nature of the magnon excitations and the extent to which the ideal gas model, which works well at higher concentrations, is useful. One aspect of the behavior near the percolation point that has recently been recognized is the fractional effective dimension, or fractal character, of the magnetic clusters. Because of this connection, studies near the percolation point may provide insight into magnetism in effectively nonintegral dimensions.

Spin Glasses

Certain disordered magnets undergo transitions to a state commonly referred to as a spin-glass state, rather than to the more familiar ferromagnetic or antiferromagnetic states. The spin-glass state, which has also been found in arrays of electric dipoles and quadrupoles, is characterized by the absence of long-range order, a property it shares with the paramagnetic phase, and by the presence of hysteresis, which is a characteristic of ferromagnetism. The appearance of the spin-glass state is signaled on the microscopic scale by a rapid decrease in the rate of fluctuations in the local field, a phenomenon sometimes referred to as spin freezing. Although the unique properties of spin glasses have been recognized for little more than a decade, they are a major topic of basic research. Originally discovered in semidilute magnetic alloys such as CuMn and AuFe, spin-glass behavior has also been observed in magnetic insulators like $Eu_xSr_{1-x}S$ (Figure 4.5) and $Cd_{1-x}Mn_xTe$.

The characteristic features of spin-glass behavior are believed to

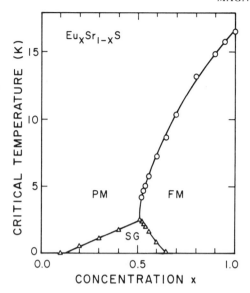

FIGURE 4.5 Phase diagram of the insulating spin glass $Eu_xSr_{1-x}S$. PM, paramagnetic; FM, ferromagnetic; SG, spin glass. [H. Maletta, J. Appl. Phys. *53*, 2185 (1982).]

arise from the presence of a large number of local minima in the free energy of the system. As the temperature is reduced, the system becomes trapped in one of these local minima. Transitions between the minima give rise to the irreversible behavior reflected in the hysteresis. The large number of minima is a consequence of a property known as frustration. Frustration refers to the absence of a unique arrangement of moments in the ground state. Unlike a ferromagnet, where the moments in the ground state are parallel, or an antiferromagnet, where there are interpenetrating lattices of oppositely directed moments, the spin glass has a large number of nearly degenerate ground states with widely differing noncollinear spin arrangements. The multiplicity of ground states can arise in a number of different ways. In the archetypal systems like CuMn and AuFe the frustration is induced by the random distribution of magnetic atoms and the oscillatory nature of the exchange interaction between them, which favors parallel or antiparallel alignment depending on the separation between the sites. In spin glasses like $Cd_{1-x}Mn_xTe$ the interactions are short ranged and favor antiparallel alignment of the moments. In this case the frustration arises from the dilution and the topology of the lattice.

Research on spin glasses focuses on understanding the onset of irreversible behavior and the properties of the ground states. Despite intense theoretical activity there appears to be no clear consensus on the nature of the spin-glass state as yet. Some analytic and simulation studies indicate that it is not a true equilibrium phase; rather, it is a metastable state analogous to that found in ordinary window glass. In contrast many experimental studies indicate behavior indistinguishable from that of a thermodynamically stable phase. It is likely that the spin-glass state is characterized by a broad range of relaxation times extending to values at least as long as the duration of the experiments. If this is the case, in an operational sense it may not be important whether the spin-glass state is truly stable.

Theoretical studies of the spin-glass transition have generally involved the analysis of infinite-range models, with the expectation that they retain some of the features of the real systems. Efforts are being made to understand the appearance of the long relaxation times at the onset of the transition. On the low-temperature side, the properties of the ground states are being analyzed along with the corresponding elementary excitations and their contribution to the specific heat and inelastic neutron scattering, for example.

Spin-glass behavior has been established in a great many materials, seemingly rivaling in number those showing conventional magnetic order. Many of the spin glasses have been studied in detail both in terms of their static behavior, as reflected in the magnetization and specific heat, and their dynamics, the latter being probed by use of magnetic resonance, inelastic neutron scattering, Mössbauer effect, and ac susceptibility, muon spin rotation, and ultrasonic measurements. Recently, two topics in the field have achieved considerable prominence. The first pertains to the study of so-called re-entrant spin glasses, which are systems that pass from the paramagnetic to ferromagnetic and then to the spin-glass phase with decreasing temperature (e.g., $Eu_xSr_{1-x}S$ for $0.52 < x < 0.65$). The issue here is whether the spin-glass state in a re-entrant spin glass is different from the spin-glass state in a system that has no intervening ferromagnetic phase. The second area is the nature of the macroscopic anisotropy in spin glasses, which is being probed in torque and electron paramagnetic resonance measurements. In this case the important question is the range of validity of various novel three-axis or triad models for the anisotropy and the low-frequency dynamics.

COMPUTER SIMULATIONS IN MAGNETISM

Few magnetic models are exactly soluble, and approximate methods of solution turn out to be either inaccurate or complex. This situation poses an increasingly difficult problem since current models of purely theoretical interest as well as those appropriate to real, physical systems are themselves relatively complex. Computer simulation studies span the gulf between theory and experiment. Often one can change the model to make it more like the physical system. One can in a controlled manner examine the effects of finite system size and surfaces, imperfections, and more complicated interactions between magnetic moments, for example (Figure 4.6). Different simulation methods have now been developed for addressing different problems in magnetism. For example, the bulk, macroscopic behavior of magnetic models and the dependence on variations with temperature and magnetic field can be determined by Monte Carlo methods. In principle, we could calculate the properties of these models in terms of properly weighted averages over all the possible microscopic states of the system. In practice, however, there are too many states to enumerate, and one is simply unable to carry out the calculation. Using various Monte Carlo methods, we can estimate the behavior of the system accurately by sampling only a small fraction of the possible configurations. Different ways have now been developed for carrying out this

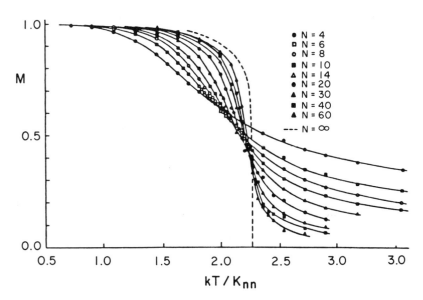

FIGURE 4.6 Monte Carlo calculations of the magnetization versus temperature for finite-size, two-dimensional Ising models. The various curves display the results from different size arrays ranging from 4 × 4 to 60 × 60. The broken curve is the exact result for the infinite lattice. [D. P. Landau, *Magnetism and Magnetic Materials*, AIP Conference Proc. No. 24 (Am. Inst. of Phys., New York, 1974), p. 304.]

sampling. The Monte Carlo method has proven successful, particularly in providing information about magnetic systems in the vicinity of their phase transitions. Simulations have been used to locate transition temperatures and to determine if the transition is first order (discontinuous) or second order (continuous).

Solutions to other problems in magnetism demand knowledge of the time dependence of the microscopic fluctuations in various magnetic models. Such behavior may be probed with very high accuracy using magnetic resonance or neutron scattering. Although we can predict the behavior of each magnetic moment for a short period of time, we find that the time development is determined by the environment (i.e., other nearby magnetic moments), which is itself changing. A computer simulation method known as spin dynamics is used to update the environment of each moment constantly, and hence determine the time development of the system as a whole. This dependence in time and space may then be analyzed using suitable Fourier transform techniques so that elementary excitations such as magnons or solitons can be detected. Most studies involve only classical models. Over the last decade, however, progress has been made in understanding various ways in which quantum lattice models may be simulated. Early work has provided information about the properties of low-dimensional XY and Heisenberg magnets.

FUTURE DEVELOPMENTS

Looking ahead to the next decade one can identify a number of areas where significant progress is expected. In particular, the utilization of various mappings to solve problems in lower-dimensional systems is one field where major advances are likely to be made. In the area of metallic systems, the behavior of intermediate valence compounds is becoming better understood as are the low-temperature properties of magnetic impurities in nonmagnetic hosts. One also looks forward with guarded optimism to continued progress in solving what may be the oldest and most difficult problem in this field: transition-metal ferromagnetism.

The study of disordered magnets is likely to become even more important, particularly since spin-glass-like behavior is being found in a rapidly growing number of materials. There is a need for a unifying picture similar to that provided by the renormalization group approach that will bring together the results obtained from a variety of measurements in different systems. In such a development it is likely computer simulations will play an important role in testing theories under controlled conditions.

5

Semiconductors

INTRODUCTION

There are only a handful of scientific or technological discoveries that have revolutionized society. Within the past few decades, none has held as central a role as the computer and communication technologies. Spectacular progress in these is directly connected to materials research in semiconductors and other materials used in electronic devices (Figure 5.1). If the rate of progress that has characterized this technology is to continue into the next decade, our scientific understanding of such subjects as semiconductor surfaces, interfaces, and defects and of deliberately structured materials—either geometrically or spatially—will be indispensable. Instead of reaching a plateau after its initial explosive growth following the discovery of the transistor based on semiconductor physics and materials, materials research related to semiconductor technology is expected to receive another impetus to further growth from the advent of very-large-scale integration (VLSI).

Alongside these exciting technological developments semiconductor physics has continued to be a surprisingly rich and fertile field of scientific inquiry. Current experimental and theoretical developments offer tantalizing suggestions that we may be able to understand some of the properties of these materials at a microscopic level. However, recently discovered new phenomena such as the quantum Hall effect or

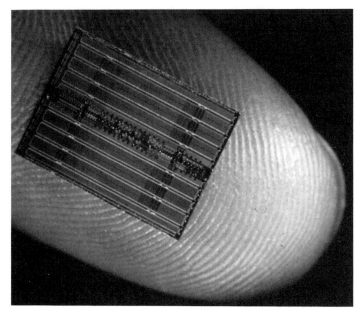

FIGURE 5.1   An experimental 1-megabit dynamic random-access silicon memory chip. The width of the smallest features is just 1 micrometer. Access time is 150 nanoseconds. (Courtesy of IBM Thomas J. Watson Research Center.)

deeper insights into the transport properties of disordered solids are constant reminders that not everything can be anticipated or claimed to be understood. The ability to prepare materials deliberately with atomic arrangements not found in nature is another reminder that science and technology are often symbiotic. The tools for preparing these materials were developed for computer technology and are now used to prepare and characterize materials that may, in the future, provide even more powerful and cost-effective computer components. In this chapter, it is not our intent to discuss technology in any detail or the essential role of semiconductor materials in it, but rather we hope to convey a brief perspective of the status in semiconductor science. However, as it is a field that is characterized by a relatively unique interplay between science and technology, it is useful in this introductory section to indicate the relationship of science, to be described in the following sections, to technological development.

Semiconductor science includes the growth and characterization of materials as well as the study of physical phenomena. Spurred by technology, the study of growth and characterization of materials will remain a competitive area. New and improved materials for applications ranging from infrared detectors, solar cells, and solid-state lasers to transistors and VLSI circuits will continue to be in great demand. The properties of surfaces of semiconductors is an active area of solid-state physics. New theoretical calculations and experimental techniques show promise of resolving many of the outstanding issues related to the structure of semiconductor surfaces. The interactions of ions, atoms, and molecules with semiconductor surfaces play a significant role in the processing of semiconductor materials, ranging from epitaxial deposition to reactive ion-beam etching. In related studies, most semiconductor device materials require contacts of one kind or another. Semiconductor-solid interfaces are therefore an important area of investigation.

The role of point, line, and planar defects in the electronic properties and yield of semiconductor devices has long been recognized, and much has been understood about such defects. However, there are still outstanding issues that need resolution, and it is expected that these will continue to interest scientists and technologists for the foreseeable future.

One of the more exciting areas of solid-state physics in the past few years has been the role of disorder and dimensionality on transport in solids. Lithographically produced structures, two-dimensional inversion layers, and high-mobility semiconductors have been widely used in the investigation of phenomena related to quantum transport localization, Coulomb interaction effects, and the integer and fractional quantum Hall effects. Heterostructures continue to be investigated for their optical and electrical properties. Recent technological developments in the production of high-speed transistors in GaAs-based epitaxial multilayers has brought renewed and increasing emphasis on this class of materials. Over the coming decade we expect these materials to be explored intensively for electronic applications requiring high speed and high-density integration.

The search for new semiconductor materials or ingenious methods for fabricating known semiconductors with the potential for novel device geometries is expected to be an active area of interest. The desire to obtain an understanding of amorphous semiconductors remains strong. Technological applications such as copying, solar cells, and optical storage will continue to drive this field.

SURFACES AND INTERFACES

Scientific interest in surfaces and interfaces of semiconductors is worldwide. The use of ultrahigh-vacuum technology over the last decade has provided data on clean surfaces, surfaces with controlled exposure to impurities, and interfaces. The availability of synchrotron radiation has made possible the generation of data on the electronic properties of semiconductors with unprecedented resolution. The recently developed scanning tunneling microscope has provided, for the first time, a direct image of the arrangement of atoms on one of the more complicated surfaces of silicon [(111) 7 × 7] (see Figure 7.1 in Chapter 7). Theoretical approaches have provided reliable estimates of the energies of semiconductor surfaces as functions of atomic positions. The latter enable one to rule out a variety of possible surface models by comparing their relative energies.

The last decade can be characterized as one in which a great variety of experimental and theoretical techniques were developed. It can also be characterized as one in which it was realized that understanding surfaces and interfaces is difficult but important problems to solve.

A combination of the results of a variety of experimental techniques using x rays, electrons, ions, and atoms has provided evidence that our understanding of the atomic arrangement at surfaces is only now beginning. In fact, it is generally agreed that only one semiconductor surface—that of GaAs (110)—is currently known reliably and accurately. A particularly encouraging development in the theory of clean surfaces has been the ability to calculate the total energy of crystals as a function of atomic geometry. Such calculations have already provided evidence (confirmed experimentally) that the notion of the buckling of surfaces—a long-held view in this field—is valid more for ionic semiconductors such as GaAs than for the covalently bonded Si. More surprisingly, it has been proposed, and current experiments support this view, that a surface of Si rearranges its atomic positions to form bonding more characteristic of carbon (pi bonding) than of Si. Parallel to the study of clean surfaces, the effects of impurities, both physisorbed and chemisorbed, have been investigated on a number of different semiconductors and their surfaces. These impurities include H, O, Cl, and F, as well as metals such as Al or Pd, on Si. The spatial location and electronic effects of impurities have been investigated both by structural studies and, for example, by vibrational high-resolution electron energy loss and infrared absorption spectroscopies.

Semiconductor-metal, semiconductor-oxide, and semiconductor-semiconductor interfaces have been intensely investigated over the last decade. With the availability of high-resolution probes and controlled environments under which such interfaces can be prepared, the number of theoretical models that can explain the properties of such interfaces has been sharply reduced. In the case of semiconductor-metal contacts, a particularly important result has been the demonstration that chemical reaction between semiconductor and metal is a dominating factor in interfacial properties. This reactivity studied at a monolayer level by atomic and electronic structural techniques has posed some fundamental questions about the formation of Schottky barriers.

DEFECTS IN SEMICONDUCTORS

Defects in semiconductors are essential for the operation of semiconductor devices as well as detrimental. They have therefore been studied for a number of decades. The properties of shallow impurities were well understood during the 1960s primarily through optical absorption experiments and effective-mass theory. In contrast, progress in understanding deep impurities and other point defects (deep centers) has been slow, largely because of technical difficulties. Most experimental techniques using bulk samples ran into problems associated with the presence of shallow impurities at concentrations greater than the defects of interest. Theoretical techniques using primarily the cluster approximation, which simulates the infinite crystal by a small number of atoms, often yielded poor results.

Major advances in both experimental and theoretical techniques for the study of deep centers occurred in the last decade. During the early and mid 1970s, a new family of experimental techniques was developed using junctions (*p-n* junctions and metal-semiconductor junctions, for example) instead of bulk samples. The advantage here is that one can use electric fields to sweep mobile carriers out of the junction region, thus effectively simulating a material without shallow impurities. A variant of these techniques, known as deep-level transient spectroscopy, is particularly powerful because individual deep levels appear as peaks on a continuous spectrum. In the past 5 years or so, we have seen the evolution of a large number of hybrid techniques. For example, optical detection of magnetic resonance combines electron spin resonance (ESR) with luminescence and is thus capable of simultaneously probing the local symmetry and the chemical identity of atoms (which is an ESR feature) and electronic energy levels (which is a luminescence feature).

Also during the past 5 years or so, a new theoretical technique has been developed, based on the mathematical tool called Green's functions, which enables one to avoid the cluster approximation and treat an isolated defect in an infinite crystal with accuracy comparable with that achieved in the study of the perfect host crystal. This has provided a detailed picture of the electronic structure of many classes of deep centers. For each deep center, the number, relative energy positions, and wave-function character of localized states can now be explained in terms of simple physical models. Trends for classes of impurities (or for the same impurity in different hosts) are better understood.

In the area of extended defects, the most noteworthy developments have been the achievement of very-high-resolution micrographs and of theoretical simulation techniques, which lead to more reliable identification of the nature of the defects. Major advances have been made in understanding the role of some extended defects in electronic devices. Most notable is the appreciation of the role of dislocations in gettering defects from the active region of devices and the similar role played by oxygen precipitates.

REDUCED DIMENSIONALITY IN SEMICONDUCTORS

Advances in semiconductor technology, especially in the silicon metal-oxide-semiconductor technology used to make the devices that are central to computer memories and other applications, have also made possible the observation and study of two-dimensional electron systems in which the electron density can easily be varied over two orders of magnitude (from about 10^{11} to about 10^{13} cm^{-2}). These systems are two dimensional in the sense that the motion of the electrons in the direction perpendicular to the semiconductor-insulator interface is constrained to a region of about 10 nm by the interface barrier and externally applied electric fields. The first clear demonstration of the two-dimensional character of these electrons was made in 1966.

Many kinds of structure have now been shown to exhibit the reduced dimensionality first seen in metal-oxide-silicon devices. They include heterojunctions (structures in which two different materials adjoin, usually epitaxially), quantum wells formed by two heterojunctions, and superlattices formed by periodic arrays of quantum wells or by periodic variations of impurity concentrations. If the confining potentials quantize the energy levels to give level spacings comparable with or greater than the thermal energy $k_B T$ and the energy level broadening, then the motion of the electrons will have a two-dimensional character. The metal-insulator-semiconductor structure has the advantage that the carrier density is easily varied by changing the voltage across the device. Although work on superlattices has concentrated on the GaAs-(Ga,Al)As system, because of the favorable growth conditions and simple band structures that they present, there is also considerable work on so-called type II superlattices, in which occupied energy levels in one material lie at the same energy as empty levels in the other.

Intensive studies of electron transport, hot-electron effects, strong and weak localization in one and two dimensions, impurity-band effects, piezoresistance, many-body effects, charge-density-wave effects, and magnetotransport effects including the quantum Hall effect have been carried out on these systems. The ability to vary the electron density has been of great help in allowing meaningful comparisons between theory and experiment to be made. Structures involving GaAs-(Ga,Al)As heterojunctions have the advantage of high mobility, especially when donor impurities are spatially separated from electrons. This can lead to electron mean-free paths of the order of 1 μm at low temperatures, more than an order of magnitude larger than the best values attained in silicon. The smaller electron effective mass in GaAs leads to larger energy splittings both for the quantum levels induced by confining fields and for the Landau levels induced by magnetic fields, which means that lower magnetic fields and higher temperatures can be used than for comparable phenomena in silicon.

The reduced dimensionality of the systems being discussed here has made possible the experimental observation of a number of important physical effects. For example, weak localization effects and the remarkable quantized Hall conductance phenomena, both discussed in Chapter 1, have been observed in these systems. The two dimensionality of these systems also leads to a situation where the electron-electron interactions make a major contribution to the electronic energy levels, as has been verified in far-infrared spectra of silicon inversion layers.

OPTICAL PROPERTIES OF COMPOUND SEMICONDUCTORS

As interesting (and important) as the optical properties of elemental semiconductors are, compound semiconductors, mainly the III-V materials ($A_{III}B_V$), add much more scope to this area of work. The III-Vs cover a wide energy-gap range (0.172-2.24 eV), are direct gap (not just indirect gap) in much of the range, possess high electron mobilities, can be made into alloys, and, above all, can be made into heterojunctions and are powerful light emitters. Thus, they have the potential to be made into light-emitting diodes (LEDs) and lasers, not to mention photodetectors (and various high-speed transistors as well). In fact, optoelectronics is totally dependent on these materials, i.e., on their optical properties. The binary crystals GaAs (the prototype), GaP, and InP have become important bulk substrate materials, and their optical properties, in their own right, are heavily studied. Also, their bulk optical properties serve as a reference for an entirely new and large area of work that is unique: III-Vs permit the construction of heterojunctions, and this in turn makes possible the construction of quantum-well heterostructures (QWHs) and superlattices (SLs), and thus deliberately designed quasi-two-dimensional structures. This achievement (i.e., quasi-two-dimensional heterostructures) of mainly the past decade puts III-V materials, and their technology, in a special category that promises to be of a revolutionary nature. Also this development is of immediate and long-range use in devices.

The two-dimensional nature of a QWH or SL breaks the crystal bulk symmetry and, for example, removes the heavy-hole, light-hole degeneracy of, let us say, GaAs of thickness smaller than 500 Å. In addition, the confined-particle states, electron and hole, partition the conduction and valence bands and permit exciton absorption (and recombination) to be observed in an abnormally large range, including (300 K, 0-10 kbar) up into the region (energy) of higher band minima (L and X). All the usual optical properties are modified by the quasi-two-dimensionality of QWHs or SLs. This, of course, is becoming an intensive area of study for a variety of III-V heterostructures, which preferably are lattice-matched (e.g., $Al_xGa_{1-x}As$-GaAs), but even in some cases can be strained-layer heterostructures (e.g., $GaAs$-$In_xGa_{1-x}As$ or $GaAs_{1-x}P_x$-GaAs). The undoped QWH or SL is of interest at low and at high carrier levels and serves, moreover, as a reference and comparison for similar heterostructures with impurities introduced into the wells or barriers, as is necessary for device applications.

In the area of optoelectronics, III-V QWHs and SLs promise to have a profound effect. Already major improvements have been effected in semiconductor laser performance. In the form of QWHs, monolithic, single-diode structures have achieved laser power levels from 100 mW to over 2 W. In these heterostructures the large asymmetry in electron-hole behavior permits a major redesign of valence photodetectors and makes possible other unique hot-electron devices. Of further interest, impurity-induced disordering can be used to convert, selectively, quantum-layer regions to bulk-layer regions, or lower gap to higher gap, thus making possible interesting device geometries (and microgeometries) and consequently integrated optical and electronic structures. There is little doubt that the optical properties of III-V materials will be a major area of study for 10 and more years and, in general, will be the basis for many further developments in optoelectronics (more sophisticated lasers, LEDs, detectors, real-space negative resistance devices, higher-speed transistors, and integrated versions of all of these processing simultaneously charge and photons). Clearly, progress in this area of work will depend on the skill and progress in III-V crystal growth and development. It is not unreasonable to assume also that QWHs and SLs will be constructed in II-VI and other semiconductor crystal systems with interesting optical properties.

AMORPHOUS SEMICONDUCTORS

Amorphous-semiconductor physics is concerned with the structural, vibrational, and electronic properties of noncrystalline semiconductors. By material, the field divides into two principal subfields: (1) tetrahedrally bonded group IV elements, mostly Si or Ge, and alloys with each other or with H and (2) the chalcogenides S, Se, or Te, alloyed with each other or with group IV or V elements. By phenomena, the field has numerous subtopics that parallel much of semiconductor physics as a whole.

Within the past decade, by far the most important discovery has been n- and p-type doping of hydrogenated group IV amorphous semiconductors, abbreviated a-C:H, a-Si:H, and a-Ge:H for hydrogenated amorphous carbon, silicon, and germanium, respectively. Related technological applications have rapidly followed, led by worldwide efforts in photovoltaics but also including demonstrated applications in xerography, vidicons, and thin-film transistors. Most of the attention for both the physics and technology is focused on a-Si:H because in this system more than in others there is the promise of studying the intrinsic disorder of a prototypical amorphous semiconductor. Because of overconstrained bonding conditions, true glasses cannot be expected with fourfold coordination. Experimentally this is seen in the form of incomplete or dangling bonds and other local structural inhomogeneities, which lead to gap states that obscure the basic semiconducting properties of interest, for example, activated conductivity, doping, and distinct band gaps. For Si, hydrogenation heals the dangling and other weak bonds and thereby permits the study of most of the previously observed effects. In the last decade there has been a burst of activity worldwide to capitalize on the scientific and technological promise of doped, hydrogenated group IV amorphous semiconductors.

The bonding between atoms in amorphous chalcogenides is different from that present in the group IV semiconductors. Hence, their atomic arrangement and the defects associated with this arrangement are also quite different. It is generally believed that electrons are paired at dangling bonds (defects) in the chalcogenides in contrast to, say, amorphous Si where the dangling bond is associated with a single charge. The chalcogens show strong photostructural changes at photon energies comparable to band gaps rather than bonding energies. For example, volume changes of several tens of percent are observed. The microscopic origins of these changes are not known. Chalcogenide materials are being explored for photoresist and optical storage applications.

While real qualitative advances have been made in the past decade, quantitative and predictive understanding of the basic phenomena associated with disorder are still lacking in both classes of amorphous semiconductors. There is considerable scientific challenge in the problems of knowing the principal sources of disorder (bond angles, intrinsic defects, and role of impurities, to name a few) and in discovering which semiconductor phenomena are unique to the disordered, amorphous state rather than remnants of analogous effects in the ordered, crystalline state. Key to the physics is the sorting out of the innumerable chemical and materials-science preparation and characterization aspects.

FUTURE PROSPECTS

It is our belief that the rate of progress in understanding phenomena and materials and in manipulating materials to obtain deliberately arranged geometrical and spatial structures will accelerate in the next decade. We expect semiconductor research to be an active area of interest not just because of technological forces but also because of our increased experimental and theoretical capabilities.

Semiconductor Surfaces and Interfaces

A variety of experimental and theoretical techniques will be applied to investigate the nature of atomic rearrangement or reconstruction on semiconductor surfaces. There will be an increasing trend toward understanding the nature of gas-surface interactions for both scientific and technological reasons. Reactions such as etching or deposition of materials with directed external radiation such as that introduced by ions or lasers are likely to be of increasing importance in semiconductor technology.

Basic research on semiconductor interfaces will grow in the coming years for the following two reasons: (1) The essential role played by semiconductor interfaces in microelectronic devices will become even more significant as the trend toward miniaturization continues; (2) experimental techniques and computational methods are available now for interfacial studies that bridge theory and experiment. Key areas of interest and progress can be identified by the following interfaces.

SEMICONDUCTOR-SEMICONDUCTOR INTERFACES

Of special interest is the low-temperature growth of p-n (or n^+-n, p^+-p) junctions. Since the diffusion distance (which increases with temperature) is an intrinsic limitation on device dimension, low-temperature processing is likely to become crucial for achieving submicrometer structures in VLSI devices. The low-temperature epitaxial growth of high-quality Si on Si with a controlled doping is a key issue. This will require understanding of the atomic process of growth of pure Si in ultrahigh vacuum, the addition of dopant, and interfacial defect formation and control during growth.

Ion implantation and laser annealing have been combined to obtain a dopant concentration in Si much higher than its equilibrium solubility. Ultrafast interfacial growth by energetic beam annealing will be a subject of continuing interest. The motion of a liquid-solid interface at high speed, its nonequilibrium nature, the heat dissipation, and the atomic mechanism involved will be subjects of study.

Interfaces in man-made superlattice structures will be another subject of study. Epitaxial growth of materials will be of particular interest, as will the growth of defect-free and stoichiometric GaAs layers for VLSI devices.

SEMICONDUCTOR-INSULATOR INTERFACES

Currently the most important semiconductor-insulator interface is the Si-SiO$_2$ interface. This is almost entirely due to the use of this combination of materials in curved electronic devices. However, other insulators on Si or GaAs will be actively explored as new device concepts are explored. More studies are required in order to understand the atomic structure, composition, and property correlation of the interface, for example, its charge-trapping states. Defect formation on Si during high-temperature oxidation is a subject of current study; it may lead to a better understanding of the intrinsic defects in Si and also of the structure and kinetic processes that determine the behavior of the interface. Laser annealing of Si on SiO_2 and thermal growth of large grains of Si on SiO_2 are subjects of technical importance.

SEMICONDUCTOR-METAL INTERFACES

The research on metal-Si interfaces has experienced rapid growth. This growth has been fostered by technological demands. We expect it to continue for several years. Also, we expect an increasing emphasis on the study of metal-compound semiconductor interfaces. The link between ultrahigh-vacuum studies and those carried out in ambient environments may bridge basic research and technological applications of these interfaces. A desire to seek fundamental understanding of the origin of Schottky barrier formation will motivate continuing research on this topic.

Defects in Semiconductors

Ion implantation is currently used to introduce controlled amounts of shallow impurities in semiconductor devices. The samples are then annealed in order to redistribute the impurities to electrically active sites. This process depends on the type of annealing used. Shallow-impurity diffusion is also affected by oxidation, the growth of a silicide, and other processing steps. The understanding of migration mechanisms both with and without thermal equilibrium is a challenging and important problem for both science and technology.

The problem of identifying deep centers is essential for a complete scientific understanding of defect processes and valuable for technology in enabling appropriate processing steps in the fabrication of devices to be chosen. The study of defect reactions under external stimulation (electron injection, temperature, and laser irradiation, for example) is only beginning, and many effects are not understood. Extended defects such as dislocations are detrimental in device performance. The electrical properties, for example, of the core structure of dislocations and the role of impurities in making them conducting are still not understood. Overall, the study of extended defects is closely related to materials processing for devices. The main problems are the understanding of the conditions under which these defects grow, their identification, migration kinetics, and role in reactions. Surface and interface defects are becoming more important as technology evolves toward the use of shallower junctions. Understanding of the pinning of the chemical potential at Schottky barriers is an outstanding problem whose resolution may involve defects. The nature of defects at the Si-SiO$_2$ interface still poses a difficult problem. Process-induced surface defects are important for technology, but current understanding is limited. The theory of surface and interface defects is primitive.

Systems of Reduced Dimensionality

Where might one expect further activity in systems of reduced dimensionality?

QUANTIZED HALL EFFECT

Work on the fractional and integer quantum Hall effect will continue, and its observation in a wider range of materials can be expected. If the precision being attained now is confirmed by additional work, the quantized Hall resistance may become a resistance standard or a secondary standard.

GROWTH TECHNIQUES AND LITHOGRAPHY

Continuing improvement in semiconductor growth techniques and in lithography can be expected to lead to use of a wider range of materials and to new device structures. In particular, it is possible to reduce dimensions to the order of 10 nm by lithography and to the order of 5 nm by shadowing techniques. This means that it is possible to construct conventional devices small enough so that electrons have a low probability of being scattered during their motion from one contact to the other and should behave ballistically.

SMALL STRUCTURES

The increased ability to fabricate small structures now makes it possible to reduce the effective carrier dimensionality even further. Narrow lines on surfaces can be expected to lead to corresponding confinement of the carriers inside the semiconductor. For conductivity processes to appear one dimensional, it is only necessary that the relevant mean-free-path parameter be large compared with the channel width. Such one-dimensional behavior has already been observed in a variety of samples. One-dimensional behavior in a quantum sense requires that the carriers be confined in a distance comparable to the electron wavelength at the Fermi surface, which is just out of reach at present but can be expected to be achieved in the coming decade. Studies of magnetic-flux quantization in small structures of normal metals are also being pursued.

HETEROSTRUCTURES

Superlattices of type II are materials in which there is energy overlap between filled states in one material and empty states in another; InAs-GaSb is the prototype. They are likely to be studied more extensively and in a wider range of materials. In these systems, electrons and holes lie in adjacent layers. This makes possible new experiments involving excitonic effects and collective effects. Strained-layer superlattices, in which the conditions on lattice-constant matching across an interface are relaxed, should also extend the range of materials and structures that can be studied. Given the ability to make small surface structures, it is possible to construct surface superlattices in which the carrier density and the strength of the potential can be varied.

THE TWO-DIMENSIONAL WIGNER CRYSTAL

The elusive two-dimensional Wigner crystal, the electron crystal expected in a degenerate low-density electron system at low temperatures, may finally be observed in inversion layers at semiconductor surfaces in the coming decade, as its classical analogue was observed in electrons on liquid helium in the past decade. Exciting new possibilities arise if electrons are placed on a thin film of liquid helium on a substrate in which a periodic- or random-potential is imposed.

6

Defects and Diffusion

INTRODUCTION

The field of defects and diffusion in solid materials is concerned with the structure of possible flaws in otherwise homogeneous materials and with their observable consequences in the properties of materials. Of particular interest are point defects, which are fairly localized on an atomic scale; line defects, such as dislocations; and boundary defects, such as surfaces, stacking faults, and grain boundaries. Although it is one of the older fields in condensed-matter physics and materials sciences, it remains an attractive arena for the observation and description of new physical phenomena and therefore maintains an enduring interest of physicists, in addition to researchers with other scientific orientations.

An indication of the vigor that defect concepts still possess is that they mold the physical understanding of many phenomena in apparently unrelated fields that are of interest in condensed-matter physics today. Several examples serve to make this point:

Solitons made their appearance in defect physics as lattice dislocations. Coupled electron-lattice complexes have long been exemplified by self-trapped holes and other polarons. These ideas have recently been combined into the unexpected form of the coupled electron-lattice

127

modes that form the soliton charge carriers that have excited interest in connection with the transport properties of conducting polymers such as polyacetylene.

Two-dimensional structures have become the focus of much recent activity for their unusual characteristics with regard to melting and phase transitions. The melting of adsorbed overlayers, or equivalently of impurity planes intercalated in layered compounds, to form a hexatic "floating raft" phase, is described in current theory by the thermally activated dissociation of dislocation dipole pairs. Similarly, the theoretical building blocks of the roughening transitions of surfaces and interfaces are just the steps and jogs of classical model surfaces.

Textures in liquid crystals, which led to dislocation descriptions of defective crystals, echo in the topical description of structure both in solids that support incommensurate charge-density waves, through discommensuration structures, and in the structural characteristics of the magnetic phases of liquid ^{3}He that exist only below 1 mK.

Local tunneling systems have long been the model for inversion of molecules and for the pocket states of centers tunneling among equivalent configurations in crystals. They return in recent advances as the central characteristic causing the linear specific heat and dynamical echo phenomena of apparently all amorphous solids (metals, ceramics, and polymers) and many disordered solids (e.g., the β aluminas).

Internal friction from stress-induced changes of defect structures, first understood for such classic systems as C in Fe and the damping of brass reeds, is now applied to the analysis of both backbone and side-chain effects in polymeric materials and even in natural carbonaceous materials such as coal and amber.

While chosen primarily for their critical importance in research fields of current interest in solid-state physics, these examples do indicate the flow of seminal ideas back and forth between areas that are clearly physics and those that are not and the way in which new subfields have emerged from areas of defect physics. Because it links directly to practical materials, the field of defects and diffusion exhibits these interconnections to a high degree, and accordingly presents the greatest problems for concise summary.

NEW FIELDS FROM OLD: AN EXAMPLE

A wide variety of choice is available to exemplify the growth of new subfields from areas of defect physics. For example, the surfaces of crystals, and the faults, reconstruction, steps, and other configurations adopted by them, now form a separate field discussed in Chapter 7.

The evolution of new vigorous subfields will be exemplified here through brief descriptions of three fertile subfields of particle-beam irradiation of materials.

Phase Microstructure and Phase Generation in Radiation Fields

In the past few years it has become apparent that when energetic radiation produces atomic displacements in solid materials it not only generates defects, defect aggregates, dislocations, and voids but may also give rise to phases that are not present without prior exposure to radiation. Here we mean phases in the thermodynamic sense of spatially bounded regions with distinct compositions and/or physical properties and with abrupt interphase boundaries. These phases are truly radiation induced, in the sense that they are thermally unstable in the absence of radiation. They thus differ from radiation-enhanced phases, which are merely thermodynamically stable phases that undergo more rapid formation when assisted by radiation-induced mixing.

Radiation-induced phases have been widely observed in binary and multicomponent alloys, including semiconductors. They occur also in insulating solids in the form of metal colloids in NaCl or amorphous islands in crystalline quartz, for example. Several possible causes for their formation can be suggested. One is that the excess point defects produced and maintained by the radiation field may favor a state or solid phase different from the one that is stable in the absence of radiation. Examples of this are the formation of amorphous silicon, of highly supersaturated crystalline solid solutions, and of the disordered state of ordered alloys, all during irradiation. These states are retained after irradiation when the thermal reordering is sufficiently slow. It is not yet known whether systems exist for which a transformation from one crystalline phase to a different phase is made energetically favorable solely because excess defects are present in either or both of the phases. An alternative mechanism involves the way that radiation-induced segregation typically redistributes the components of a solid on a microstructural scale of 10^{-3} to 10 μm. There are two causes for this redistribution: first, persistent defect fluxes are set up during irradiation, and, second, certain components couple preferentially to the defect fluxes. Changes of composition then shift the system locally into a region of the phase diagram that differs from that occupied by the overall homogeneous alloy. A new radiation-modified equilibrium phase may then precipitate locally, or an existing equilibrium phase may dissolve.

Surface and Near-Surface Probes

Recent years have seen the development of important methods for the microstructural and microchemical analysis of the surface and near-surface constitution of solid materials. Particle-beam methods have so revolutionized analytical science in these areas that the first 10,000 Å (1 μm) of a material structure can now be analyzed with a chemical sensitivity often approaching 1 part in 10^7, with a depth resolution of 100 Å, or with a lateral resolution of 0.1 μm or better. Different techniques are complementary in depth, spatial, or chemical resolution; experts have therefore learned to use an arsenal of powerful new particle-beam methods to obtain detailed near-surface chemical analysis of materials structures.

Ions with energy greater than 100 eV incident on a crystal penetrate the surface and dissipate their energy and momentum. By detecting the x rays caused by the collisions it is possible to detect trace impurities at the level of 1 in 10^8 in favorable cases. Collisions also eject surface atoms from the material as secondary atoms or charged ions. By detecting the surface chemical species as the material is sputtered away it is then possible to determine the original microchemistry of the material. The detection can be performed by secondary-ion mass spectrometry (SIMS), in which the secondary species are fed into an isotope-imaging mass spectrometer, or by Auger spectroscopy of core levels excited by an auxiliary electron beam. The resolution limit perpendicular to the surface is ~100 Å. Lateral resolution is limited at present to about 1 μm in SIMS and 0.05 μm in Auger probes. In chemical sensitivity SIMS can often achieve a level of 1 part in 10^6 or better. By Auger methods the sensitivity is reduced to 1 part in 10^3, but the depth resolution may be improved to a few atomic layers since the Auger electrons from deeper layers are scattered and lost. Sputtering then allows a three-dimensional chemical map of the surface region to be acquired.

The mixing that takes place through the top 100 Å or so while atoms are being sputtered away causes a steady-state nonuniform subsurface concentration profile to develop even in a chemically uniform bulk material.

In a second category of technique, the incident and detected fluxes involve the same species. The incident beam, usually H^+ or He^+, impinges on the surface. Part is reflected by Rutherford backscattering from atoms in the subsurface region and with an energy transfer that depends first on the target mass and second on the depth to which the particle penetrates. The spectrum of reflected particle energies contains information about both depth and chemical structure that can be separated to obtain accurate chemical information with a depth resolution of about 100 Å. Alternatively, channeling methods can be used to probe the position of an atomic species in the lattice structure. When the beam is directed along a crystallographic axis so that the particle range is long, atoms located off crystal lattice sites scatter particles into other channels and into the bulk material. In this way, the location of off-site atoms can be determined with some precision. It is also possible to measure the misfit at the heterojunctions between different crystals by this method.

Ion-Beam Microfabrication

As described in Appendix C, new types of crystals, compounds, alloys, microstructures, and other materials can bring with them formerly unimagined opportunities for scientific and technological advancement. This is particularly true of microfabrication, in which components are chemically and structurally tailored for particular application on a microscopic scale. Several important methods are based principally on ion-beam methods. Here we mention three areas of major effort:

Ion implantation is the process in which foreign dopant or alloying ions are accelerated to energies of typically 1-300 keV and implanted into the near-surface regions of target materials. Implantation depths are typically of the order of 1 μm, depending on the implantation energy, and the profiles are reasonably well understood. The compositions achieved by implantation are not constrained by usual thermodynamic or kinetic limitations. As ions penetrate the solid they create lattice displacements, and the material is damaged by the implanting beam. Further manipulation of the resulting nonequilibrium structure is then often desirable.

Ion-beam mixing often involves a surface layer, typically a few hundred angstroms thick, of one material being mixed with a bulk substrate of a second material by means of a penetrating ion beam. The advantage of this procedure is that higher concentrations of new alloy phases may be achieved using only a small fraction of the irradiation fluence normally needed for ion implantation. The effects of sputtering, cascade mixing, radiation-induced segregation, and radiation-enhanced diffusion remain important but are as yet imperfectly understood.

Lithography has been used in the semiconductor industry for many years. It proceeds by damaging a chemically resistive material using a photon, electron, or ion beam and then preferentially etching away either the damaged or undamaged region and the substrate beneath it. Complex patterns with resolution of about 0.1 μm may be inscribed into semiconductors for device fabrication by these methods.

CALCULATIONS OF DEFECT STRUCTURE

A decade ago no reliable procedures were available by which accurate calculations of cohesion could be carried out for most solids. The exceptions to this statement were highly ionic solids and, to some degree, simple metals. It is now possible to make calculations that correctly indicate the small differences of relative energy among alternative bulk crystal structures of metals and covalent compounds. The energy of a surface and its electronic structure can be calculated quite well.

A mainstay of defect calculations for two decades has been the modeling of crystal configurations using energies derived from model interactions among atoms. A large crystallite with appropriate boundary conditions is used. These methods have played a major role in the development of fundamental ideas about defect structure in simple materials. The practice for metallic systems has been to sum pairwise potentials, and this has some measure of validity although the model clearly lacks a rigorous basis. Nevertheless, qualitatively useful studies even of such complicated defects as impurities bound in mixed-dumbbell interstitials have been forthcoming. For ionic materials, elaborate codes have been developed to add appropriate treatments of core polarizability to the coulombic and core-core interactions, in order to simulate the total energy of a configuration more accurately. By whatever method, the energy as a function of configuration is finally computed, and the result provides the input for calculations of molecular dynamics and properties of relaxed point defects or surfaces, for example.

If not highly precise, these methods can nevertheless often reproduce systematic trends in data such as *F*-center excitation energies and Schottky pair energies with absolute values within 20 percent of those observed. Long experience, the systematic elimination of errors, and fine tuning of the codes have made the procedures reliable for ionic materials such as alkali halide and alkaline earth fluorides, which have large excitation energies and hence stable polarizabilities. Advances over the past decade include reasonable description of defect volumes (errors formerly led to large discrepancies) and entropies. A good understanding has been achieved for important model impurity problems involving dopants, both with and without effective charges, in halide and fluoride structures. The case of rare-gas impurity properties in these lattices warrants special mention. Ongoing efforts seek to broaden applications of this general approach to less-ionic materials such as oxides, where added effects of covalency must be simulated.

A powerful approach to the properties of metallic and covalent systems, mentioned in Chapter 1, is provided by methods derived from the density functional formalism. Normally this involves iteration of the one-electron Green's function to obtain a self-consistent electron density and hence the energy by integration over position. Various paths to these results employ frozen cores, pseudopotentials, or other strategies to avoid the atomic core. In applications to total energies of extended solids these methods have been highly successful: relative energies, and hence stabilities, of different structures are predicted with a precision that often is better than 0.1 eV. Note that the problem of calculating the total energy of the electron liquid from first principles is circumvented rather than solved in these approaches. Variational and Green's function Monte Carlo approaches to the specific problem of the electron liquid appear to offer feasible future routes.

The difficulty in representing excited configurations, particularly those containing inhomogeneities such as charge localization, probably places serious limitations on the applications of the density functional method to areas of excited-state spectroscopy. At the time of writing, applications to the ground state of dilute impurity systems, both in metals and semiconductors, and including transition metal centers, have been completed to provide good insight into the local structure. Typically the approach is to use a small cluster consisting of the atom and its neighbors as a perturbation on the electronic structure of the perfect solid; this is then iterated through to self-consistency and the properties derived. Magnetic systems are treated using an *Ansatz* for the spin-polarized density functional. The impurity ground state is discussed more successfully than the excitation energy, particularly for deep levels. Vacancies have been treated as relatively simple defect centers in both semiconducting and metallic crystals. The problem of incorporating lattice relaxation into the calculation must be solved before the energies obtained can usefully be compared with experiment.

An alternative approach, which has developed mostly in connection with insulators, is now finding important applications to metals and covalent materials. This is the unrestricted Hartree-Fock approximation. A critical recent breakthrough has been the development of many-body perturbation theory to correct the Fock results for correlation by an expansion in pair excitations. Almost 90 percent of the correlation energy is returned by these methods in many problems, so practical calculations can be completed with chemical accuracy, or 0.1 eV.

One important advantage of the Hartree-Fock approach is its ability to deal accurately with excited configurations. This is particularly the case for excited configurations that differ in symmetry from the ground state, so that the two remain unconnected by pair excitations in the many-body perturbation theory.

FUNDAMENTALS OF ATOMIC MOBILITY

Until recently, computer simulation using molecular dynamics has provided the sole method by which jump rates can be calculated quantitatively for a given defect in a model crystal in which atoms interact through a specified potential function. This method can be employed for model crystallites of 10^2-10^3 atoms, using periodic boundary conditions that minimize surface effects. Computer runs of 10^4-10^5 iterative steps necessarily involve only $\sim 10^3$ vibrational periods in order that the iteration can mimic smooth dynamics. Therefore, the method has been useful only when a number of jump events take place in 10^3 periods. This has limited investigations to fast-diffusing species and to high-temperature properties. Topics on which attention has focused include diffusion in liquids, in superionic conductors, and on surfaces. Each of these will be mentioned further in what follows.

It should be emphasized that existing evaluations of statistical theories describing atomic jump rates have disagreed with the results of correct molecular-dynamics calculations for the identical model systems. This has been a consequence of the insufficient accuracy in the statistical evaluation afforded, for example, by absolute rate theory (ART). For the jump frequencies encountered in real crystals it has been possible to perform dynamical simulations only at the highest temperatures. Consequently there has been an almost complete absence of accurate theoretical information for ordinary materials about the way atomic jump processes are determined by the potential energy of interaction among the atoms. Much discussion has arisen over the past decade about the possible dependence of the energies, entropies, and volumes of migration on temperature and pressure, since they are sensitive to theoretically intractable derivatives of the jump rate with respect to these variables. The meaning of the isotope effect and its dependence on thermodynamic coordinates have remained equally obscure.

One recent advance in this area is to correct ART predictions for those nonrandom return jumps that are inherent in the dynamical system and that therefore cause an intrinsic error in the rate theory formulation. It turns out that in most crystals these amount to only 10 percent of the jumps, even at high temperature, so the correction is small. Apparently absolute rate theory alone is a relatively sound first approximation in many cases.

ART is nevertheless valid only for a limited range of jump problems. A second formulation, Brownian rate theory (BRT), is potentially useful under different circumstances from those for ART. In ART it is supposed that the system randomizes completely once a jump occurs; therefore ART requires correction for subsequent dynamical events that occur before randomization. In BRT the inertia of the jumping system is retained but introduces random motion into the remaining system, which causes a viscosity. No clear synthesis of these two approaches has appeared. The physical sense of BRT becomes apparent for problems such as adatom diffusion on a smooth surface where the adatom may move many lattice spacings before undergoing a significant collision with the lattice. Under these circumstances the assumption in ART of immediate randomization is incorrect. Diffusion occurs instead by long flights broken by collisions in a way that could conceivably be described by viscosity. No first principles evaluation of the damping has, however, been possible as yet for atomic migration.

One striking application of the BRT has been to the breakaway of dislocations in a stress field. Here, thermal activation through pinning points plays a role, as does the inertia of the dislocation at its terminal velocity in the viscous field. An attractive feature of the phenomenon is that the viscous drag from the electronic system can be modified to an observable extent by the superconducting transition, which thus affects mechanical properties. For this example the model holds together in a semiquantitative way.

The influence of quantum constraints on atomic jump processes has been the source of a large theoretical literature that has dwarfed the incidence of verifiable experimental reports citing observations of quantum effects. Well-established observations exist for tunneling of even quite heavy atoms among pocket states of asymmetric defect configurations; these results include, for example, off-center Cu substituted in salts and, more recently, Zn-Al mixed dumbbell interstitials in Al. These systems undergo tunneling transitions among equivalent configurations and have unmistakable signatures of symmetry and tunnel splitting of the energy-level structure. An elegant system that still lacks adequate model treatment is the motion of exceedingly heavy self-interstitials in certain metals. In Pb, for example, the interstitial is observably mobile below 1 K, so it may in fact delocalize rapidly in the perfect crystal at 0 K. A possible complication in metals, not currently well understood, is coupling to the conduction electron excitation continuum, with excitation density characteristically proportional to excitation energy in the Fermi liquid.

In the 1970s, the migration of light impurity atoms through crystals was modeled using polaron concepts to find regimes of multiphonon (thermally activated) hopping at high temperature, power-law few-phonon hopping at intermediate temperature, and (perhaps attainable) propagation at the lowest temperatures. Feynman path integral methods are now thought to present a viable route for more accurate calculations of transition rates, particularly in combination with Monte Carlo treatments of the classical lattice modes. Extensive studies of the most promising systems, principally H interstitials in bcc refractory metals, have seemed consistent with this modeling to some workers, although this is not fully accepted. Certainly, large isotope effects are measurable in mechanical relaxation and specific heat, for example. The situation is complicated by H trapping in tunneling levels at other, heavier interstitials and also at dislocations. True migration of a light particle at low temperatures in these materials has proved elusive.

Quantum crystals offer the best opportunities for examination of atomic mobility modified by quantum-mechanical requirements. Solid H_2 and D_2 are complicated by rotational transitions, and Ne, Ar, and CH_4 are too heavy; the focus therefore falls almost entirely on the isotopes ^{3}He and ^{4}He. It is possible that exchange delocalizes vacancies in these materials and mixes them into the crystal ground state so that they are never absent in equilibrium, even at 0 K. Explicit measurements have revealed that the vacancy content of hcp ^{4}He and bcc ^{3}He is below 1 part in 10^4 as $T \to 0$. The defect density is nevertheless sufficient to promote diffusion with a diffusion constant $D \sim 10^{-6}$ cm^2 s^{-1}, characteristic of the melting temperature of more ordinary solids. It is further observed that the vacancy formation energy is equal to the activation energy for diffusion over a wide range of volumes in bcc ^{3}He. Evidently this structure has little or no barrier to migration, although the hcp structure at lower molar volume does exhibit activated hopping. With the use of variational and Green's function Monte Carlo methods with model pairwise forces it is currently possible to reproduce cohesive and structural properties of the heliums quite accurately. The possibility therefore exists that theory may take the leading role in the exploration of these point-defect properties.

Quantum crystals may also have unusual extended defects. For example, the liquid-solid interface shows exaggerated lateral mobility in certain circumstances. Also, melting waves are observed at perturbed liquid-solid interfaces, owing to the high heat conduction permitted by the superfluid. Finally there is an expectation that dislocations may become delocalized. Measurements nevertheless show that the string model of dislocations, pinned at vacancies, describes the mechanical behavior down to temperatures of 1.5 K for solid ^{4}He and to about 0.2 K for solid ^{3}He. For both point and extended defects, therefore, the promise of remarkable defect behavior remains mostly to be verified in future work.

COMMENTS ON ACTIVE AREAS

What follows are brief descriptions of additional areas that seem particularly noteworthy in the light of past developments or potential future interest. More material concerning surfaces and interface properties will be found in Chapter 7.

Point Defects in Simple Solids

Steady progress has been made over the past decade in collecting and interpreting values of the formation and migration properties of simple defects in prototypical solids. In some cases the known activation energies, typically ~1 eV, have changed by less than 10 percent from values quoted two decades ago. The intervening years have nevertheless been filled with careful effort. In many cases, activation energies for self-diffusion and for vacancy formation have been measured over large temperature ranges using new techniques. Curvatures of Arrhenius plots are now commonplace; the interpretation in terms of the temperature dependence of mechanisms, hopping parameters, or defect clustering nevertheless remains a difficult and unsettled area. The assistance of definitive theories will probably be required before these subtleties are resolved. To convey the difficulty, one may note that for many bcc metals the Arrhenius plot for diffusion has been known for many years to exhibit strong curvature, yet the cause remains imperfectly resolved. One reasonable possibility is the anomalous behavior of phonons in these metals.

Persistent efforts are necessary in this area. The correct characterization of defect processes in simple metals, salts, and valence solids is essential if progress toward a predictive framework for complex materials of interest for engineering applications is eventually to be forthcoming. In this connection the realm of definitive information is now advancing from noble metals and alkali halides to refractory and other metals, covalent solids, and refractory oxides. This must be recognized as a major achievement of the field.

An indicator of future progress is the healthy arsenal of techniques now applied to quantitative characterization of defect structure in metals. In the early 1970s, for example, the existence of the self-interstitial in Cu as a dumbbell configuration was established using ultrasonic attenuation, neutron scattering, and diffuse x-ray scattering. To these techniques have been added positron annihilation, muon spin rotation, perturbed angular correlation spectroscopy, various special NMR techniques, ion channeling, the Mössbauer effect, and other specialized probes. These advance far beyond the resistivity measurements and occasional specific heat and Bragg x-ray scattering experiments available before 1970, and the prospects for steady future progress are improved accordingly.

Surface Diffusion

Only at temperatures typically below half the melting temperature does the diffusion of atoms on clean surfaces usually resemble the site-to-site hopping of atoms in the bulk crystal. Most available information for lower temperatures pertains to refractory metals that can be cleaned in ultrahigh vacuum to secure reproducible results. Field ion microscopy provides a microscopic probe of adatom mobility and clustering in the low-mobility regime, and fluctuation spectroscopy of adatom Auger signals is a second, more macroscopic, probe. Reasonable Arrhenius behavior has been observed in a number of experiments covering limited temperature ranges; evidence for quantum behavior at low temperatures has recently been reported for H and D adsorbates. Phase transitions of the bulk and surface reconstructions are both expected to change the diffusion characteristics, but as yet these have not been investigated. Anisotropic diffusion is seen to occur, and interesting mechanisms have been deduced, for anisotropic surfaces containing ridges, for example. These have been simulated in computer dynamical calculations.

At higher temperatures, simulations indicate fast surface diffusion. Moreover, so many mobile point defects and surface ledges are activated that the surface becomes quite rough and is subject to rapid fluctuations. Interestingly enough, the surface layer itself is observed in simulations to exhibit liquidlike behavior below the melting point of the solid. These novel dynamics are, of course, confined to a skin on the crystal surface since the bulk cannot melt. None of these high-temperature processes has yet been observed for real crystals.

Photochemical Processes

It has been known for several decades that optical interband transitions in salts can create point defects. In alkali halides the products are H centers (a negative halogen molecule at an anion site) and F centers (an electron trapped at an anion vacancy). The exciton created in the optical event self-traps, and a nonradiative decay channel leads to the defect production. A number of similar processes warrant mention here. For example, the Jahn-Teller instability of the vacancy in Si transforms electronic recombination energy into the motion required to surmount the barrier to atomic migration; thus, excitation promotes fast migration. Another example is dislocation glide, which occurs at high excitation levels in solid-state quantum-well lasers made, for example, from GaAs and $Ga_{1-x}Al_xAs$. Yet a further example is photodesorption, in which the surface of a crystal absorbs a photon and an ion is subsequently ejected from the surface. It is believed that Auger processes convert what was formerly a negative ion into a positive ion, which then desorbs under the repulsive field of its neighbors. Angle-resolved effects from molecules oriented on the surface may be expected; site symmetries have been elucidated from the paths of desorbing ions.

Photochemical processes of this type offer new opportunities for investigation. Interesting progress has been made in recent years using picosecond pulse-probe laser techniques to monitor the decay of excitons into F and H centers. It appears that a V_k center forms after an unresolvably short time and that it evolves into the H center. Nowadays, laser pulses can be created in the 10-fs (10^{-14} s) time domain, which is faster than most lattice vibrations. A special opportunity therefore exists in the future to examine photochemical processes, including photon-induced point-defect migration, on the time scale of the atomic jump process itself.

Molecular Dynamics

Computer simulation of dynamical processes in solids has led to vivid insight into complex mechanisms, to the discovery of qualitatively new processes, and to quantitative mimicking of processes that occur in real solids. As mentioned above, molecular dynamics has served as the only source of numerous insights into the dynamics and stability of defects, for example in surface structure and diffusion.

More recent successes that warrant mention here include the accurate treatment of ionic motion in fast ion conductors such as AgI, in which the Ag sublattice disorders, and in CaF_2, where the F sublattice undergoes large fluctuations involving defects and mobility. Model interatomic forces have reproduced observed diffusion rates quantitatively. Also of major interest are dynamical studies of the early stages of precipitation. In these investigations the embryo is found to nucleate from the disordered state, possessing from the earliest stages the appropriate symmetry as a precursor of the eventual lattice structure. It is not clear what other approaches could possibly provide such direct access to these important phenomena.

The future of computer simulation and dynamics contains research problems to match whatever complexity new generations of computers can handle. Direct modeling of the mechanical properties introduced by dislocations may require consideration of $\geq 10^6$ particles in place of today's 10^3. Bulk mechanical behavior, including grain-boundary structure, may require still more. Future dynamics programs may be coupled to fast quantum chemistry routines to replace pair forces by more realistic solid-state modeling of the crystalline potential energy. Surface treatments, including particle-beam mixing, the resulting nonequilibrium structure in alloys, and the influence of directed heat input by laser processing, all appear readily susceptible to investigation by simulation. It seems highly probable that applications of this type will ensure a significant future role for computer studies of defect properties.

Dislocation Motion in Glasses

The glide of dislocations in crystals has long been recognized as a determining factor in plastic behavior. It has recently been recognized that many of the same effects occur in glasses also, despite the fact that the geometrical characteristics are somewhat less clear owing to the amorphous structure of the solid. The experimental fact is that slip bands are observed after plastic flow in certain metallic glasses. They occur as expected on planes defined by the maximum shear stress and, in general, resemble similar processes in crystals. Computer modeling of dislocations introduced into a Lennard-Jones glass shows that the core structure and the long-range elastic field remain stable after the atoms are allowed to relax. By way of comparison, a vacancy in the

Lennard-Jones glass disappears into the structure when relaxation is permitted; however, bond models with forces that vary with bond angle can lead to stable vacancies.

These initial discoveries establish that glassy materials can, in some cases, support defect structures resembling those in crystals. The area warrants further effort in the future to determine the range of phenomena that occur and the way in which such defects move.

Defect Imaging at Atomic Resolution

A notable development over the last decade has been the refinement of experimental methods that can image crystals and defect structures with spatial resolution at the atomic level. These are not universally applicable probes but instead generally require particular sample characteristics for successful detection of atomic locations and defect geometry. They are nevertheless extremely powerful techniques when handled well. Two such probes that are surface sensitive are the scanning vacuum tunneling microscope and the field ion microscope. These are described in Chapter 7. A third, the high-resolution electron microscope, can image defects on the surface or within the bulk of the crystal.

Over the period from 1970 to 1982 the resolution of commercial transmission electron microscopes improved from 4 to about 1.5 Å. Since the beam passes through the entire sample, which must therefore be quite thin (~1000 Å thick), these methods are naturally adapted to linear and planar defects aligned with the beam. Vivid patterns of atomic distributions in the perfect crystal can be obtained for appropriate thin films. Dislocation and defect structures at properly aligned grain boundaries can also be imaged. Early high-resolution successes concerned the planar defect structures of stacking faults and polytypes and the defects accommodating nonstoichiometry in oxides. More recent applications involve grain-boundary structures and the geometry of heterojunctions, for example of semiconducting materials in device configurations (Figure 6.1). In all cases the apparently clear imaging of atomic positions is at least partly illusory; careful theoretical modeling is required to obtain precise interpretation of the relevant diffraction processes. Most of the necessary theoretical machinery is now widely available.

The past decade has also seen the parallel development of scanning transmission electron microscopes, which form an electron probe focused to about 3 Å. Images with atomic resolution are then formed by monitoring such properties as scattered intensity as the beam is

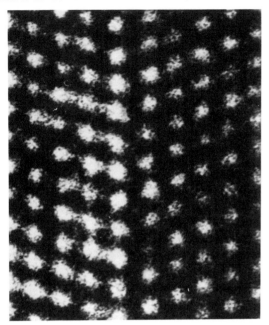

FIGURE 6.1   Atomic resolution image of the interface (lines) between Si (left) in the (110) projection and epitaxial $NiSi_2$ (right), which has the fluorite structure. Each dark blob is the image of two projected rows of atoms in a sample about 100 Å thick. (Courtesy of J. C. H. Spence, University of Arizona.)

rastered in a suitable pattern. Heavy atoms located on carbon films have been imaged individually by these means; other direct uses at atomic resolution include the analysis of small precipitates; chemical analysis at high resolution can be completed by energy-loss methods.

SOME DIRECTIONS FOR FUTURE RESEARCH

A qualitative understanding of phase microstructure and phase generation in radiation fields is developing, but detailed model descriptions and even the basic theoretical framework remain largely to be developed. One phenomenon, radiation-induced homogeneous precipitation in undersaturated solid solutions, has been described using a simplified quasi-thermodynamic theory. This is a research area in its infancy. It adds a new dimension to the currently active field of phase relations, phase transformations, and the stability of phases.

At present there exist no precise methods by which the steady-state nonuniform substrate concentration profiles that develop while atoms are being sputtered away from the solid surfaces can be predicted quantitatively, so accurate analysis awaits improved understanding of the damage, displacement, and diffusion processes in the subsurface region.

The structures and energies of point defects are apparently coming within the grasp of *ab initio* theoretical calculations. Indeed, electronic aspects of calculations for point defects can already be treated comprehensively, but the rather complicated problem of lattice relaxation and its effect on the electronic system is not generally tractable at present.

As with density functional methods, the new capabilities provided by the use of the unrestricted Hartree-Fock approximation, corrected by the use of many-body perturbation theory, open a wide range of problems to future exploration, including chemical pathways, equilibrium solid-state lattice geometries, and defect properties, although most investigations to date involve surface and adsorbate problems. It is probable that, over the next few years, accurate research on spectroscopic applications in both pure and defective solids, but particularly when some degree of electronic localization occurs, will make use of this approach. For example, *ab initio* calculations on the F and F_A center excitations of salts are apparently yielding excellent predictions. Recent results of cluster calculation on metals indicate that local electronic excitations of defects may be predictable to within an uncertainty of 0.1 eV. This accuracy is adequate for most practical purposes, so the opportunities for new applications in the future appear particularly inviting.

It is fairly clear that the discrepancies between the results of statistical theories and those of molecular-dynamics calculations for atomic jump rates in systems in which diffusion is occurring will be resolved over the next 5 years for model crystals with reasonably simple potential functions. Detailed properties of atomic jumps in model crystals, including isotope effects and thermodynamic derivatives of jump rates, will thus become accessible for the first time. Attention will then focus on the ability of the assumed potential functions to simulate that of the real crystal, much as for defect formation.

For the future it seems assured that atomic resolution investigations of defect structure in solids will continue to build momentum. Applications of these methods to surface studies in ultrahigh vacuum currently remain in their infancy but hold extraordinary promise. The real-time recording of solid-state reactions at atomic resolution under controlled conditions can be expected to reveal a wealth of detailed mechanisms over the next decade. Therefore seminal contributions may be expected in areas such as precipitation and phase transitions, where the atomic mechanisms and motions at the reaction front are of central interest.

7

Surfaces and Interfaces

INTRODUCTION

The outermost layer of atoms in the surface of a crystal has been studied for over five decades, since the observation of diffraction of electrons by the two-dimensional array of atoms in the surface of a nickel crystal established the wave nature of the electron in a clear and unambiguous fashion. For years, the field was plagued by the inability to prepare surfaces sufficiently clean and well characterized to ensure the reproducibility of data. This problem was solved by the development of ultrahigh-vacuum techniques during the past two decades. Now, when a surface is prepared, it can be maintained perfectly clean for 1 hour or longer, while a variety of precise measurements are performed on it.

In the past 10 years many new experimental probes have been used for the study of the structural and dynamical properties of the crystal surface and of atoms or molecules adsorbed onto it. These new probes are summarized in Table 7.1. In parallel with these developments, there have been rapid advances in surface-physics theory. Numerous examples now exist where important new conclusions have followed from the interplay between theory and experiment. In addition, we see substantial progress in the development of both *ab initio* descriptions of the electronic structure of clean and adsorbate-covered surfaces and of the dynamics of crystal surfaces.

TABLE 7.1   Experimental Techniques Used in the Study of Physical Properties of Surfaces and Interfaces

Experimental Technique	Structure	Elementary Excitations on the Surface	Atoms and Molecules on the Surface	Interface Between Solids and Dense Media
Ion beams	X			
Raman spectroscopy		X	X	X
Scanning vacuum tunneling microscope	X			
Synchrotron radiation	X		X	
Electron energy-loss spectroscopy	X	X		
Electron microscopy	X			
Atom/surface scattering	X	X	X	
Low-energy electron diffraction	X		X	
Neutron scattering	X	X	X	
X rays	X			
Infrared spectroscopy		X	X	X
Spin-dependent electron scattering		X		
Brillouin scattering		X		X
Diffusion of adsorbed species			X	
Molecular beams			X	
Laser-induced desorption or fluorescence			X	
Inelastic electron-tunneling spectroscopy				X

This chapter is concerned primarily with the physics of the outermost atomic layer or two of single crystals in an ultrahigh-vacuum environment, along with that of monolayer quantities of adsorbate atoms or molecules upon it. The adsorbates may be chemisorbed, i.e., bound to the surface tightly via chemical bonds similar to those encountered in molecules, or physisorbed, where only a much weaker van der Waals attraction traps the adsorbate near the surface. Also, a substantial interest exists in the microscopic nature of the solid-gas or the solid-liquid interface, where the first few layers in the low-density phase may have properties modified profoundly by their proximity to the solid interface. Some of the new methods of studying the interface between a crystal and vacuum are directly applicable to the analysis of the liquid-solid or the gas-solid interface, as we shall see.

If one considers the outermost atomic layer of a perfect crystal, one may inquire whether it is a replica of a plane of bulk atoms or whether it differs importantly. The latter is frequently the case. The atoms may shift off the sites expected from the bulk structure to form a new, low-symmetry phase unique to the surface. This is known as surface reconstruction and is observed on many surfaces. The electronic structure of the surface may be unique, because of unsaturated dangling bonds. Such electronic states are often responsible for the particular chemical reactivity of the crystal surface.

Adsorbate overlayers are a rich area for study. The environment of a chemisorbed species differs greatly from that in an isolated molecule, and this leads to new electronic configurations. At finite coverage, adsorbates may interact directly via the overlap of their wave functions, indirectly via the perturbation of the electronic structure of the substrate, or by local strains induced by chemisorption. Such lateral interactions control ordering of the overlayers, and thus control the thermodynamic phase diagram of the adsorbate/substrate combination.

Physisorbed rare-gas atoms are weakly bound to the surface and may move parallel to the surface relatively unhindered. One may view the substrate as a passive entity whose role is to confine the atoms to a plane parallel to its surface; hence the absorbate system constitutes a realization of two-dimensional matter. From recent theory we know that physics in two dimensions differs profoundly from that in three dimensions. Hence, study of physisorbed overlayers offers insight into basic issues of statistical mechanics in two dimensions.

We have outlined why, from the point of view of fundamental physics, the study of surfaces is of great interest. In addition, advances in our understanding of surface physics have a direct impact on other areas of science. In the chemical industry, solid-state catalysts are extremely important in many manufacturing processes. While a particular catalyst often proves highly efficient for a limited range of reactions, little is understood about the origin of this specificity. Knowledge of the basis for efficient catalytic activity will allow the design of new catalytic structures. Practical catalysts are typically complex, multicomponent systems, prepared in powder form, and operating in a high-pressure environment; they differ substantially from a single-crystal surface, prepared and cleaned in ultrahigh vacuum. However, the fundamental principles elucidated in the study of the single-crystal surface and its interaction with atoms or molecules will form the basis of a deeper understanding of how real catalysts function. Also, diagnostic methods developed in surface physics have been applied to the analysis of real catalysts. In the past 3 years, a combination of surface-physics methods (Auger spectroscopy, low-energy electron diffraction, and electron energy-loss spectroscopy) has been used to study the decomposition of hydrocarbons on single-crystal platinum surfaces. This has allowed us to trace out, in a step-by-step fashion, how their decomposition leads to formation of a carbonaceous layer, and the consequent poisoning of the surface as a catalyst.

In materials science, there is great interest in new multilayered structures formed by deposition of two or more materials and constituting a macroscopic material with unique properties. In idealized form, such a structure consists of single-crystal films, with the thickness of each controlled and in the range of a few tens to a few hundreds of angstroms. Semiconductor superlattices are one example of such materials. There are also new superlattices formed from metals or from combinations of (ferromagnetic) metals and semiconductors. The study of the single-crystal surface, and of adsorbates on it, may be viewed as the study of the first atomic layer of a new constituent in a superlattice structure. Thus, research in surface physics has a direct impact on this exciting new area of materials science.

THE STRUCTURE OF THE CRYSTAL SURFACE

If one is to understand properties of, and bonding to, the outermost layer of a crystal, a first step is the elucidation of the geometrical arrangement of the constituents. Thus, considerable effort is spent on the development of probes that provide structural data and on theoretical descriptions of the interaction of the probe with the surface.

To extract structural information from the data is formidable. The problem is that the probe must either reflect off the outermost layer or backreflect after penetrating only a small number of layers. This means that it interacts strongly with the crystal. Hence full utilization of information in the data requires a sophisticated theory that treats the strongly interacting probe completely. (In contrast, in the study of bulk structures, the probing quanta—x rays or neutrons, for example—travel long distances in the crystal; their interaction with any one constituent is weak. This leads to simple theories for interpreting data.)

It proves difficult to determine a surface structure unambiguously from one set of data. Consequently, several methods are often used to study a single structure. There is no single probe or method, such as x-ray scattering used in bulk studies, that solves surface-structure problems.

Many surface structure studies employ either low-energy electron beams, with energy in the range of a few to a few hundred electron volts, or neutral atom or ion beams. The electrons sample a small number of layers, three or four in number, while neutral atom beams sample only the outermost contours of the electron charge density. Both have a deBroglie wavelength comparable to lattice spacings, or bond lengths, and thus serve as sensitive probes of microscopic aspects of surface geometry.

Of the electron spectroscopies, low-energy electron diffraction (LEED) has been a mainstay for many years. Through its use one may identify systematic structural trends. Recently it has been established that on most metal surfaces there is a contraction between the first and second layer, with the more open faces contracted the greatest. Observations such as these have provided a major stimulus to theory.

A major development of the past decade is the utilization of synchrotrons as intense sources of electromagnetic radiation with a continuous spectrum of wavelengths, which extends from the visible, through the ultraviolet, and into the x-ray region. Many new surface-sensitive spectroscopies based on these sources have emerged during the past decade. Among them photoemission has developed rapidly. Here a photon, which penetrates many atomic layers, will eject an electron from the crystal. One measures the total current emitted by the crystal, the energy distribution of the emitted electrons, or their angular distribution. Sensitivity to the surface arises because the mean free path of the excited electron is a few lattice constants. Thus, only those excited close to the surface emerge.

Ultraviolet photons excite electrons from the valence orbitals of the atomic constituents of crystals. One finds here surface electronic states revealed by photoemission; these states are markedly affected by the surface geometry, so their spectroscopy provides important information on surface structure. A decade ago, we had few data in hand in this area, but now the influence of a surface on the electronic structure of materials has been explored experimentally for many semiconductors and many metals.

Other spectroscopies employ synchrotron radiation as the basic probe. A photon may eject a core electron from an adsorbed atom or molecule. The emitted electron wave propagates to the detector, at the same time that a backscattered portion reflects off the surrounding structure to interfere with the direct wave. Study of the energy and angular variation of the cross section for this process provides information on the local environment of the atom or molecule involved. X-ray absorption edges exhibit fine structure, with a closely related origin. Since the synchrotron's output is in the form of a broad, continuous band of radiation, these features may be explored in detail.

New areas of synchrotron-radiation-based spectroscopy are in early stages of development. For example, while x rays have not generally been considered a surface-sensitive probe, at glancing incidence their electromagnetic fields are evanescent in the substrate. Thus, the Bragg beams reflected from the crystal contain information about the outermost region of the crystal. In particular, if the surface reconstructs, new Bragg beams are induced by the reconstruction process. These beams provide information about the magnitude of the atomic displacements parallel to the surface. Recently, more complete analyses have provided information on the vertical displacement of atoms. The method may be unique in providing access to displacements associated with reconstruction, without recourse to the theories required in electron- or atom-beam studies.

Glancing-incidence neutron spectroscopy also offers the possibility of surface sensitivity, though the method has yet to be implemented fully.

One exciting new probe has appeared recently: the tunneling electron microscope. This device operates on the basis of a fundamentally new principle. If two metals are placed ten or so angstroms apart, with a potential difference between them, an electron may transfer from one to the other via quantum-mechanical tunneling. One metal is a sharp tip, while the second is the sample to be studied and is nominally flat. The current that flows is sensitive to the distance between the tip and the sample; if the tip is scanned across the sample, the current will fluctuate in magnitude in response to protrusions on the surface that change the distance from the probe tip to the sample. Features in the surface profile that influence the current are steps or defects on a length scale of a few to a few tens of angstroms, or possibly the bumps in the electron density produced by the individual atomic constituents. A direct, real space map of the surface geometry is produced.

The current spatial resolution of the device renders large-scale surface features, such as steps, readily visible. Since steps and other defects on the surface play a major role in catalysis, in the nucleation of reconstructed phases, and in other surface processes, this is an exciting development. The spatial resolution allows studies of atomic arrangements in open surface structures. Thus, the 7×7 reconstruction of the (111) surface of silicon has been probed directly (Figure 7.1), as has the (110) surface of gold.

As remarked earlier, low-energy atoms have a de Broglie wavelength comparable to crystalline lattice constants, and such atoms are backscattered from the outermost portions of the electron charge-density contours. Recently there has been a great improvement in our ability to prepare monoenergetic beams of light atoms, such as He. At

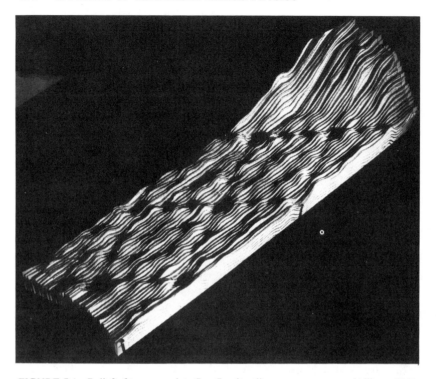

FIGURE 7.1   Relief of two complete 7 × 7 unit cells on a reconstructed silicon (111) surface, with nine minima and twelve maxima each, taken at 300°C. Heights are enhanced by 55 percent; the hill at the right grows to a maximal height of 15 Å. The (211) direction points from right to left, along the diagonal. (Courtesy of H. Rohrer, IBM Research Laboratory, Zurich.)

present, beams with energy spreads well below 0.1 meV are readily available. The study of elastic- or Bragg-scattered beams is rich, with scattering resonances evident that are associated with bound states of the atom/surface interaction potential. These are sensitive to the details of the potential, including its variation in the two dimensions parallel to the sample surface.

New high-energy ion backscattering studies place important constraints on surface geometry and provide quantitative information on whether surface atoms are shifted off high symmetry sites or if there is a contraction or expansion of the distance between outermost layers of the crystal. Here classical trajectory analyses are sufficient to analyze the data. A set of qualitative concepts based on the shadowing of interior atoms by those in the surface has developed that may be used to interpret data. This is an experimental technique that will continue to be developed in the coming years.

Electron microscopy is another technique emerging as a surface probe of substantial importance. Transmission of electrons through thin films offers the possibility of studying electron diffraction under conditions where a single scattering description is appropriate. A dark-field imaging method can pick out only superlattice reflections and thus can be used to examine the dynamics of formation of the (7 × 7) structure on the Si (111) surface. It is even possible to observe nucleation of the new phase near steps on the surface with this technique. Electron microscopy will become an important tool for exploring a range of issues such as the nature of defects on the surface, dynamical aspects of the surface environment, and also microscopic aspects of surface structure.

A final probe that is being used extensively in surface studies is the field ion microscope. It employs a needle-shaped specimen with an electric field at its tip so strong that inert gas atoms are ionized; the ions subsequently follow the electrostatic field to an imaging screen. Since the fields are strongest at the surface and are sensitive to surface features on an atomic scale, the pattern of ion impacts on the screen images atomic details of the sample tip. Surface structure, atomic diffusion on surfaces (Figure 7.2), and chemical groupings of adatoms are the types of surface properties that are now investigated in a direct way by field ion microscopy. In addition, however, by means of strong field pulses it is possible to strip away successive surface atoms and reveal underlying structure. This makes depth profiling a feasible, if tedious, procedure. When coupled with a time-of-flight mass spectrometer the field ion microscope thus permits a structural and chemical map of the tip surface and underlayers to be built up sequentially at atomic resolution.

The past decade has clearly been one in which classical methods of deducing surface geometry have advanced in a qualitative manner, while a number of new and potentially powerful methods of analysis are under active development.

SPECTROSCOPY AND ELEMENTARY EXCITATIONS ON THE SURFACE

A major advance in understanding the physics of bulk solids occurred when inelastic neutron scattering became widespread. The neutron may scatter inelastically off of any elementary excitation such as a spin wave or phonon, and an analysis of the kinematics of the

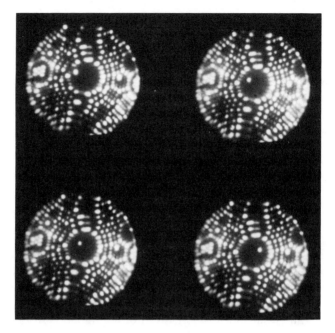

FIGURE 7.2   Movement of a single Re adatom diffusing on the central (211) plane (dark circular area) of a tungsten crystal field ion microscope tip. The four images taken at successive times show the Re adatom progressively displaced. (Courtesy of K. Stolt and G. Ehrlich, University of Illinois.)

scattering event allows one to map out their dispersion curves. Since the de Broglie wavelength of thermal neutrons is of the order of a lattice constant, one may study the dispersion relations throughout the entire Brillouin zone. Such data have led to a qualitative expansion in our understanding of perfect and imperfect crystals.

During the past 2 years, two methods—inelastic scattering of helium atoms and electron energy-loss spectroscopy—have been developed to the point where they can now be used for detailed measurements of surface phonon dispersion curves.

In the previous section we noted that we now have in hand highly monoenergetic beams of slow, neutral He atoms. These not only allow one to study fine detail in the elastic cross section but have led to the realization of high-resolution inelastic scattering studies of surface phonons. We now have data available for several insulating and metallic surfaces.

Shortly before 1970, electron energy-loss spectroscopy was used to study high-frequency vibrational motions of light adsorbates, and in

1970 the method was used to study vibrational spectra of clean surfaces. This technique has developed into one of the standard working tools of surface science during the past decade. As the surface atoms or adsorbates vibrate, oscillating electric dipole moments associated with the motion lead to intense inelastic scattering of the electron through small angles. Virtually all of the experiments study these near-specular losses. In this configuration one studies only modes whose wave vector lies close to the center of the corresponding two-dimensional Brillouin zone.

During the past 5 years, the first studies of electrons that suffer inelastic scattering through large angles have appeared. A selection rule that applies to near-specular scattering breaks down in the large-angle studies, with the consequence that many more modes can be explored. The first experiments explored the vibrational spectrum of adsorbates. When off-specular spectra are combined with the near-specular data, a rather complete picture of the adsorbate geometry emerges.

Within the past year off-specular electron energy-loss studies have been employed to obtain surface phonon dispersion curves for a clean surface and also for a surface covered with an ordered adsorbate layer. We thus have two methods that may be utilized to explore surface phonon dispersion curves, and the coming years should prove to be an exciting era in the spectroscopy of surface vibrations. It should be remarked that the two methods discussed above will surely emerge as complementary approaches to the problem.

One area of surface spectroscopy in an early stage of development, with great future promise, is the spin-dependent scattering of electrons from a surface. One may now produce beams of spin-polarized electrons, through the use of GaAs emitters appropriately pumped with laser radiation. With these beams, elastic scattering data are obtained from ferromagnetic substrates. One may align the spin of the beam electrons either parallel or antiparallel to the substrate magnetization and detect the difference in scattering intensity. It is believed that this difference is in essence proportional to the magnetization in the crystal surface. The new data have been used to infer the temperature variation of the magnetization in a ferromagnet below the bulk Curie temperature and to provide the first results for this quantity close to the Curie temperature.

Neither inelastic atom/surface scattering nor electron energy-loss studies have provided information on the linewidths of the various modes that have been probed. It is unlikely that the electron energy-loss method will achieve sufficient resolution in the near future to generate such information, and while atom/surface studies may be able to realize sufficient resolution to measure linewidths in favorable cases, this has not been done as yet. However, both infrared and Raman spectroscopy offer high resolution, when applied to the study of bulk excitation spectra of condensed matter, and one may look toward either of these as possible high-resolution probes of surface vibrations. Both suffer from weak signals, a disadvantage if the aim is to study monolayer or submonolayer quantities of material on the surface of a single crystal. Despite this basic difficulty, recent years have seen substantial progress in each area.

There have been impressive new methods in infrared studies of adsorbate vibrations. In one version of this method the frequency of a surface electromagnetic wave (surface polariton) is swept through a vibrational resonance of the adsorbate and the resulting attenuation studied. One may achieve an appreciable enhancement here over the signal level expected in a one-bounce reflection experiment. A second, multiple-reflection, method in which the sample serves as a waveguide has been utilized to obtain beautiful high-resolution spectra of submonolayer quantities of hydrogen on a silicon surface. Finally, the thermal emission of a warm sample placed in a cryogenic environment has been employed to explore the vibrational motion of CO on a single-crystal Ni surface. In all these cases, the data yield the linewidth of the normal mode explored. These experiments employ state-of-the-art equipment of a sophisticated nature. Since infrared spectroscopy is at the moment the only experimental method that has provided data on the intrinsic linewidths of simple adsorbates on single-crystal surfaces, further application of these new approaches to a broader class of systems is of paramount importance.

Great excitement was generated a few years ago by the discovery of giant Raman signals from adsorbed molecules. The effect was discovered by electrochemists, who were exploring the interface between an electrolytic liquid and a silver electrode. The Raman cross section per adsorbed molecule was found to be larger than that observed in the gas or liquid phase by a factor of 10^5 to 10^6. This enhancement, if present for a wide range of adsorbate/substrate combinations, would render Raman spectroscopy a viable high-resolution probe of surface vibrations.

An essential element necessary for the giant signals is the presence of roughness on the silver surface. The physical picture that emerges is that such roughness couples the incident photon to surface electronic resonances, whose origin may be the protrusions themselves. Excitation of surface resonances enhance the electromagnetic field of the incoming photon, and since the Raman cross section scales as the fourth power of the field strength, a modest enhancement in the field leads to a large enhancement in the signal realized.

Since roughness plays an essential role in generating the giant Raman intensities, quite clearly this form of spectroscopy is difficult to apply to high-quality single-crystal surfaces. Also, the surface resonances must lie in the visible, and the damping that they experience must be modest. These two requirements limit the effect to only a limited number of substrates, with silver a particularly favorable material. Systematic control of the enhancement effect can be achieved utilizing a surface on which a diffraction grating is present. The resulting enhanced fields can be exploited to enhance a wide variety of linear and nonlinear optical interactions near surfaces and interfaces, in addition to the Raman effect. For example, the intensity of second-harmonic radiation obtained from an illuminated metal surface can be enhanced by a factor of 10^4, through use of surface roughness.

It is quite possible to detect Raman signals from adsorbed molecules in the absence of the giant enhancement discussed above. While the signals are weak, spectra from monolayer quantities of molecules adsorbed on high-quality, single-crystal surfaces have been observed. Raman, like infrared, spectroscopy can be carried out in a high-pressure environment or can be used to explore the liquid-solid interface, if the liquid is transparent. Thus, development of either tool to the point where it may be used to explore a diverse range of systems will be an important step.

A third optical spectroscopic technique that has experienced substantial development is Brillouin scattering of light from surface phonons, surface spin waves, and other surface excitations. One may now explore backscattering of light from metals where the skin depth is only 150 Å. Kinematical considerations restrict its application to waves with very long wavelengths (typically a few thousand angstroms), so the technique is not a microscopic probe of the surface. However, one can measure elastic and spin-wave stiffness constants in thin films and explore the influence of a surface on the acoustic or magnetic response of the medium.

INTERACTIONS OF ATOMS AND MOLECULES ON THE SURFACE

The understanding of the physical origin of various features in the atom/surface interaction potential will clarify many aspects of surface chemistry. One source of quantitative information on the atom/surface interaction potential is the analysis of the intensities of atoms scattered elastically from crystal surfaces. As mentioned earlier, recent advances allow the preparation of highly monoenergetic, highly collimated atom beams. The resulting improvement in the quality of scattering data is truly impressive, and the theory of atom scattering from a rigid, possibly deeply corrugated, surface has undergone considerable development in the same period of time. A consequence is that semiempirical interaction potentials have been constructed that describe the interaction of rare-gas atoms with several surfaces.

There are two distinct limiting cases in the discussion of rare-gas atom/surface interactions. A surface of an insulating or semiconducting material is highly corrugated, and one must understand in detail the magnitude of this corrugation whose influence on many aspects of atom motion on or near the surface is a crucial consideration. On the other hand, the low-index surfaces of simple metals are smooth, so the rare-gas atoms are trapped on the surface, but they are free to move parallel to the surface impeded only modestly by the corrugation. The physics of overlayers adsorbed at finite coverage is then dominated by lateral interactions between the adsorbate atoms.

In recent years, much has been learned of the physical nature of these lateral interactions. It is now clear that the van der Waals interactions are influenced importantly by the close proximity of the atoms to the surface. Much attention has been devoted to the phase diagram of monolayer or near-monolayer quantities of rare-gas adsorbates, in the limit where the influence of the corrugations in the substrate potential may be ignored. The phase diagram of chemisorbed systems has also been studied both experimentally and theoretically, and the strengths of the various lateral interactions deduced from it.

Information on the atom/surface potential can be obtained from experiments other than scattering experiments. The primary mechanism for the diffusion of atoms over a surface is thermal activation over the barrier between the initial and final site. This activation energy provides information about saddle points in the potential energy surface. Computer simulations of such dynamical processes are important to carry out, since they can test whether a given semiempirical atom/surface potential provides results that fit the data.

The interaction of molecular beams with surfaces has been studied actively and will yield important results in the near future. Small diatomic molecules can impact the surface, and a new element in such scattering studies is the presence of the rotational and vibrational degrees of freedom in the molecule. The molecule may now scatter off the surface with a change of rotational quantum number, and in fact the molecule-surface interaction potential itself may depend on the rotational quantum state of an incoming molecule or of one adsorbed on the surface. Recent experiments that study the scattering of H_2 molecules from an Ag surface show that the positions of the fine-structure resonances, and hence the bound-state energies of the molecules, depend on the rotational quantum number.

Lasers have also found applications in the study of various properties of atoms and molecules on solid surfaces. In the past few years, nonlinear optical techniques have been exploited to probe surfaces and interfaces. Second-harmonic generation in reflection from a surface has been shown to have enough sensitivity to detect submonolayers of atomic and molecular adsorbates. The technique is versatile. It can yield information about the dynamics of molecular adsorption and desorption, the changing of adsorption sites, the spectrum of adsorbed molecules, and the arrangement and orientation of adsorbates, for example. That it can be extended to the infrared region makes high-resolution vibrational spectroscopy of adsorbed molecules also a possibility. Several other laser surface probes have also been developed recently.

One may learn much about the kinetics of molecules on the surface by analyzing the population of various rotational and vibrational states of species desorbed or scattered from the surface; here laser-induced desorption or fluorescence and photoacoustic spectroscopy may be useful probes. When one combines such information with theory generated by computer simulations based on molecular-dynamics routines, one gains considerable insight into those aspects of the dynamics of molecule/surface interactions crucial to the understanding of chemical interactions on surfaces. One may monitor the rotational and vibrational statistics of species that come off the surface. A non-Boltzmann distribution is frequently found, and there has been considerable success in the comparison between such data and theory based on computer simulations.

THE INTERFACE BETWEEN SOLIDS AND DENSE MEDIA

In many instances, one is interested in surfaces placed in a high-pressure environment or in the interface between a surface and a dense medium such as a liquid. Catalysts operate in a high-pressure environment. In electrolytic cells, a dense fluid overlies the interface.

In such systems, many experimental methods used in ultrahigh-vacuum environments are inapplicable. Any method based on the use of electron beams, or of atoms or ions, fails because the mean free path of these entities in the dense medium above the substrate is too short; the same is true of photoemission spectroscopies because the mean free path of the photoemitted electron is too short for the electron to emerge from the dense medium above the solid. Thus, many of the methods discussed above cannot be applied to the systems discussed in the opening paragraph of the present section.

However, new methods may be applied to these systems. If the dense medium is transparent, then any photon spectroscopy may be used to explore the interface, provided the signal from the species of interest (at the interface) may be detected, in the presence of the possibly large background from the dense medium. If one seeks a species not present in the liquid, then the feature of interest may lie in a frequency domain removed from that dominated by the dense medium. In these situations, techniques such as infrared or Raman spectroscopy may be employed.

If the solid substrate is transparent, then the interface may be probed by bringing the probe beam into the interface through the substrate. If it strikes the interface at an angle of incidence greater than that required for total internal reflection, the optical wave field is evanescent in the liquid, so the backscattered or reflected radiation provides information only on the near vicinity of the interface. In Raman or Brillouin studies with visible radiation as a probe, one may examine the first 200 Å of the liquid by this means. Signals may be enhanced by either fabricating the substrate into a waveguide, and employing an integrated-optics geometry, or by overcoating the waveguide by an evaporated film and coupling the incident and scattered photons to surface polaritons in the film. Some early experiments using these techniques show them to be promising for future surface studies.

In an earlier section we mentioned that, under certain circumstances, the Raman signal from adsorbed molecules may be enhanced enormously over that appropriate to the gas phase. However, since surface roughness combined with the presence of long-lived surface resonances is necessary to realize these large signals, the effect is of limited utility for the study of single crystals in high vacuum. Fortunately, just these conditions are realized in electrolytic cell environments, where, in fact, this remarkable phenomenon was discovered. The spectra so obtained are impressive: the Raman signals from the adsorbed molecules are so strong that they are comparable with those produced by the solution itself. Since roughness is present on electrolytic cell electrodes as a consequence of cycling the applied voltage, Raman scattering will serve as a powerful probe of the interface in these systems.

Practical catalysts employ the active material in the form of small particles suspended on a substrate. These high surface-to-volume systems may be probed with infrared spectroscopy; indeed, the earliest infrared spectra of adsorbed species were obtained on such systems. A new method of vibrational spectroscopy applicable to them has been developed in the last decade. This is inelastic-electron-tunneling spectroscopy. It has been known for many years that if an electron tunnels from one metal to another, through an insulating barrier between them (the insulating barrier is commonly the oxide of one of the constituents), then features in the I-V curves of the structure are produced at voltages that correspond to the energies of vibrational modes of molecular species trapped in the oxide. Analysis of these features allows one to deduce the nature of the molecular entities trapped in the oxide. The method has now been applied to systems that mimic practical catalysts, to obtain the vibrational spectra of the adsorbed species. The resolution of the method is high, if the data are taken at low temperatures (4 K is sufficient).

Thus, while many of the experimental methods of surface physics rely on the use of particle beams with constituents that have a short mean free path in matter (this is why such beams are useful for probing the surface, of course), so that they are not applicable to the study of the interface between a solid and any dense medium, new spectroscopies are being developed that are directly applicable. The latter will surely continue to evolve in the coming years and will provide an important supplement to more traditional methods.

THEORY

Advances in experimental techniques for the study of structural and dynamical properties of crystal surfaces, particularly through the use of external probes such as electrons and rare-gas atoms, require parallel theoretical efforts for their interpretation owing to the strong interaction of these probes with the system being studied. The experimental advances of the past decade described in the preceding four sections have been accompanied by significant theoretical achievements.

The theory of atom/surface scattering has made significant advances in recent years. It is now possible to calculate the intensities of the diffracted beams fairly accurately, given the potential. Calculations based on model potentials can be brought into impressive agreement with data, and a detailed understanding of the physical origin of the scattering resonances often observed has emerged.

A fundamental problem in the interpretation of atom/surface scattering data for the determination of surface structures is to relate the interaction potential to the actual nuclear positions in the crystal. It has recently been shown that to a good approximation the repulsive part of this potential is proportional to the electron charge density in the surface layers of the crystal and is consequently short ranged, while the attractive part is of the van der Waals type and is long ranged. *Ab initio* calculations now can generate electron-charge-density contours 3-4 Å outside the surface, which is as close as a helium atom incident on a crystal approaches its surface.

During the past decade, there has been a major advance in our understanding of the electronic structure of surfaces. From early theoretical studies, and simple pictures based on chemical intuition, it was clear that under a variety of circumstances one should find two-dimensional bands of electronic states localized on the surface. The development of the technique of angle-resolved photoemission has allowed the direct study of the surface state bands on a wide variety of clean and adsorbate-covered surfaces. During this period, largely through application of the density functional formalism, theorists have carried out self-consistent studies of the electronic structure of surfaces. Agreement between theory and experiment can allow one to draw firm conclusions about the structure of the clean surface or the bonding sites of adsorbates.

In the largest number of such theoretical calculations a structure for the surface is assumed, and a self-consistent calculation is then carried out of the one-electron energy states, with the nuclei held fixed. The results are then placed alongside the data, and, if necessary, the calculations are repeated for several different surface geometries until a match between theory and experiment is achieved. In a major development theorists are now actively engaged in calculations of the total energy of the surface structure, within a framework that allows the nuclear positions to be varied. Then one may seek the configuration of lowest energy to predict the surface structure of a given material.

OPPORTUNITIES

The achievements of the past decade provide some indications of the areas in which research in surface physics will be carried out in the next decade.

Surface Brillouin spectroscopy, the emergence of Raman spectroscopy as a surface-sensitive probe, and the use of field enhancement in a variety of optical interactions near surfaces and interfaces constitute a new field of experimental endeavor with substantial promise. A factor-of-2 improvement in the spatial resolution of the scanning vacuum tunneling microscope will be a major advance; this device will offer surface physics a new probe that will greatly expand our understanding of a variety of features of surface geometry.

The technique of second-harmonic generation of laser light can be used for *in situ* measurements at interfaces between two condensed media with a picosecond time resolution. This opens up many interesting and exciting possibilities for surface studies in, for example, high-pressure catalysis, electrochemistry, photolithography, and even biophysics.

In spin-dependent scattering of electrons from a surface we have a new surface probe that for the first time can be used to probe magnetism in the outermost atomic layers of crystals. For example, antiferromagnets should be readily studied by this technique, since new Bragg beams will appear if the surface orders in such a manner that the appropriate two-dimensional unit cell increases in size. One may also envision inelastic scattering of spin-polarized beams where a spin wave rather than a surface phonon is responsible for the loss. Since we have no information on the behavior of surface spin-correlation functions in the near vicinity of a bulk phase transition, there would be great interest in the study of the diffuse background to such scattering produced by spin fluctuations, particularly near a magnetic phase transition.

However, there has been little theoretical attention paid to calculations of the magnitude and the energy and angular dependences of cross sections associated with spin-dependent electron scattering. Such analyses should prove helpful, by elucidating optimum scattering geometries.

There are other questions that need to be addressed by theorists. LEED data show that on most faces of metal single crystals there is a contraction in the spacing between the first and second layer, with the greatest contraction occurring for the more open faces. These results are in disagreement with the predictions of simple pair-potential models of the crystal. Sophisticated theory is required to provide a framework for interpreting the data. Thus, as our ability to carry through *ab initio* calculations of surface structure improves, the information will be of direct value to LEED theorists and others engaged in other studies of electron spectroscopy of the surface region. The implications are exciting; LEED theorists will have in hand clear theoretical guidance when an obvious choice of structure fails to fit the data. The effort will lead to more reliable potentials for integration into a variety of analyses of the interaction of electron probes with the surface. It will also be possible to calculate, in an *ab initio* fashion, the force constants that enter models of surface lattice dynamics.

A complete understanding of the photoemission process requires knowledge not only of the electron energy states, their wave functions, and the sensitivity of both to surface structure, but it also requires knowledge of the electromagnetic field of the incoming photon in the near vicinity of the surface. This is an area where further theoretical understanding is both required and will prove fundamental not only to photoemission spectroscopy but to other surface spectroscopies addressed in this report.

In a great deal of the theoretical work on the interaction between an atom and a crystalline substrate the latter is treated as if it is a perfectly rigid structure, whose only role is to provide an effective potential that influences the motion of the atom. In fact, if an atom is placed in an adsorption site, there is a distortion of the lattice in its near vicinity. Also, if the atom is adsorbed on a particular site and it hops to a neighboring site, there will be a local distortion of the lattice that will be dragged along with it, during diffusion on the surface. As an atom approaches a crystal, to reflect off it in a scattering experiment, there will be a local distortion in the near vicinity of the impact site. As the atom recoils, a substantial fraction of its energy may be transferred to the lattice. This whole sequence of phenomena requires for its elucidation a description of the interaction of the atom with the vibrational quanta of the substrate (phonons) and a theory that provides a valid description of the consequences of this coupling. Our ability to describe atom/phonon interactions, and to exploit their consequences, is at a primitive stage at present, yet these couplings may play a crucial role in many aspects of the atom/surface interaction. Computer simulations may prove useful here.

So far, we have discussed only the scattering of atoms off the surface or their motion on it, under circumstances where the electronic configuration of the atom remains unchanged during the interaction process. For rare-gas atoms interacting with the surface, this picture is surely sufficient in most circumstances. However, when atoms (or ions) with lower ionization potentials or electron affinities strike the surface, it is possible for an electron to be transferred from the atom to the surface or for the incoming particle to pick up an electron. A rigorous description of these processes poses a real challenge. At present, classical trajectory analyses form the basis for most theories, but a fully quantum-mechanical description of the atom motion may be required.

Our discussion of the last two topics has focused primarily on the need for further development of the theory. The absence of predictive theories limits our ability to appreciate the full significance of existing data, and if definitive developments occur in the theory, surely the direction of experimental research will be affected positively. Another area where the absence of theory limits our ability to appreciate the full significance of data is electron- or photon-stimulated desorption, in which adsorbed atoms are detached from a surface on excitation by incident electrons or photons.

Our knowledge of the origin and even the magnitude of lateral interactions, particularly in chemisorbed systems, is sketchy. Since these interactions play a key role in stabilizing the various surface phases encountered in chemisorbed systems, and control the degree of short-range order present in a disordered overlayer, a more complete understanding of the underlying physics that controls their strength and magnitude is important to have.

The study of the interaction of small molecules with surfaces, with emphasis on the interchange of vibrational or rotational energy, is expected to be a lively and active area in the coming years. As our understanding of the interaction of atoms with surfaces becomes increasingly quantitative, we acquire a base upon which a clear understanding of molecule/surface interactions may be based.

As one moves from small, simple diatomic molecules to more complex entities such as hydrocarbons, we approach issues of direct interest to surface chemists. There is at present a rather limited amount of structural data on the adsorption geometry of hydrocarbons and hydrocarbon fragments. The complexity and variety of adsorption geometries possible for these systems renders a full quantitative interpretation of the data difficult, but we are seeing the beginnings of active research in the area, with attention to quantitative results.

Several laboratories are actively exploring the use of time-resolved methods to probe the kinetics of molecule/surface interactions in real time. Such data could lead to a qualitative expansion in our understanding of the kinetics involved, by direct observation rather than indirect inference. It is possible to envisage the construction of apparatus that can resolve surface kinetics at the submillisecond level, with microsecond resolution as a lower bound owing to limitations on one's ability to chop a molecular beam. Activated rate processes are easily slowed down by cooling the sample, so millisecond-resolution experiments will suffice to provide a major step forward in our understanding of surface kinetics. This is an area, virtually unexplored at present, that should prove exciting in the coming years.

8

Low-Temperature Physics

DEFINITION OF SUBFIELD

There are two central themes that define the field of low-temperature physics. One is the study of the collective behavior in the motion of quantum-mechanical fluids, such as the electrons in superconductors and liquid helium. The other is the development of technology to go to lower temperatures. Any experiment attempted at the limit of this technology is usually classified as low-temperature physics.

QUANTUM FLUIDS

The term quantum fluid is used as the generic expression to cover any material in which the particles of interest do not solidify when in the ground state. Quantum fluids remain as a liquid or gas at $T = 0$ K because the particles are sufficiently light that the ground-state kinetic energy is larger than the interparticle potential energy that normally causes crystallization.

Interest in such materials is frequently centered on how they satisfy the third law of thermodynamics. The entropy, which is a measure of the disorder in a system, must go to zero as the temperature goes to zero. Other substances achieve the condition of perfect order through spatial arrangement, usually in a lattice structure. In quantum fluids the

164

order takes place in the motion of the particles. Superfluidity, the ability to flow without resistance, is the manifestation of such order. The particles are coordinated to move in a coherent way. A single wave function can serve to describe the behavior of an entire mole of atoms, nuclei, or electrons, if they are in a fluid state at $T = 0$ K.

Superconductivity, discovered in 1911, was the first example found of such order. It describes the ability of electrical currents to flow without resistance in certain conductors. The superfluidity of liquid ^{4}He was discovered in 1937, 30 years after it was first liquified, and manifests itself in such properties as the ability to flow through capillaries of such small diameters that ordinary fluids cannot pass through them.

Basic research in both superconductivity and superfluid ^{4}He remains active. Rapid progress in the understanding of superconductivity followed the development of the microscopic theory by Bardeen, Cooper, and Schrieffer in 1957. The BCS theory attributes the phenomenon to an attractive interaction between pairs of electrons in a spin singlet (antiparallel spins) state caused by the lattice vibrations of the solid. The theory was so successful that superconductivity has now become an important tool for studies of the electron interactions in many different types of metals. In the case of ^{4}He, a microscopic theory of the interactions that lead to superfluidity has still not appeared nearly 50 years after the phenomenon first came under study. Modern research into the fundamental questions related to superfluidity has frequently been directed toward changing the nature of the transition by producing it in thin films on planar surfaces and a variety of packed powder geometries.

In the past decade there has also been much work in which superfluidity and superconductivity have been applied to entirely new types of questions. Thus, liquid ^{4}He is now widely used as a model substance for fundamental investigations in other research areas. It is a simple material with few defects and no impurities. For example, it is being used in tests of recent ideas about the onset of disorder in fluid flow. In the case of superconductivity, commercially significant technologies that depend on it are likely to become commonplace in the near future. Most hospitals will have superconducting magnets for use in nuclear magnetic resonance (NMR) tomography, and devices based on the Josephson effect will be widely used.

In a significant development, new areas of research into quantum fluids have recently emerged, to complement the knowledge that we have gained about motional order in superconductors and superfluid ^{4}He. Liquid ^{3}He has been found to undergo a phase transition similar to that occurring in superconductors, and an active search is under way for similar phenomena in other low-temperature fluids, such as spin-aligned atomic hydrogen.

Superfluid ^{3}He

A high point of the research of the last decade was the discovery of the superfluidity of ^{3}He at temperatures below 3 mK. This is the only new superfluid to be discovered in nearly 50 years. The possible existence of a paired state of atoms in liquid ^{3}He was predicted as early as the late 1950s. By 1970, the usual estimate of the transition temperature T_c for a state like that in superconductors was in the range between 10^{-6} and 10^{-9} K, well below that possible for the cryogenic technology of the day. Thus, it came as a rather dramatic surprise in 1971 when the superfluid transition was accidentally found in a series of experiments in which cryogenic methods for cooling solid ^{3}He were being developed.

The reason that the transition occurs at such an unexpectedly high temperature is that the nature of the pairing interaction in ^{3}He is quite different from that found in superconductors known to date. In ^{3}He the bound pairs of atoms form in a triplet magnetic state (parallel spins) with total spin $S = 1$ (so that $S_z = -1, 0, +1$), whereas the pairs in superconductors and in ^{4}He are in the singlet state (antiparallel spins) with total spin $S = 0$. In the former case the orbital part of the pair wave function is required to be antisymmetric in the interchange of the coordinates of the two atoms; in the latter it must be symmetric. The members of a pair having parallel spins must be in states of odd angular momentum, and in the case of ^{3}He they are in a state with $L = 1$. As a consequence of this the order parameter of ^{3}He is not a simple scalar function of position, as it is in ^{4}He, but is a much more complex entity that can undergo a variety of different distortions. In liquid ^{3}He the magnetic interaction between pair constituents is almost ten times larger than that between the paired electrons in a superconductor.

Unlike the cases of superfluid ^{4}He and superconductivity, where only one phase is found at low temperatures, liquid ^{3}He has three distinctly different low-temperature phases, called the A, A_1, and B phases. The A_1 phase exists only in the presence of a magnetic field. The boundaries between these phases as functions of temperature, pressure, and magnetic field are shown in Figure 8.1. Even at the low temperatures indicated, ^{3}He remains a fluid until a pressure greater than about 34 bar is applied.

The three phases have quite different experimental properties. Regardless of which superfluid phase is entered from the normal fluid,

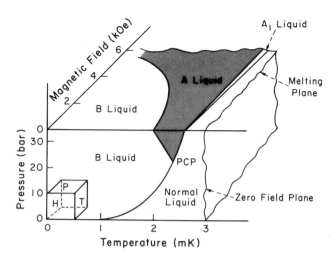

FIGURE 8.1 The phase diagram of liquid ³He.

however, a second-order phase transition like that in superconductors is observed. However, the transitions between the B phase and the A phase is a first-order transition with a minute latent heat, and the transition between the A and A_1 phases is also a first-order transition.

The interpretation of the phase diagram is that the three superfluid phases correspond to the separate onset of the pairing transition for each of the three spin polarizations of pairs in the triplet state (see Table 8.1). The B phase contains all three polarizations: $S_z = +1$, 0, and -1. The A phase contains $S_z = +1$ and -1; and the A_1 phase contains only $S_z = +1$.

An important consequence of the odd angular momentum of the orbital wave function is that superfluid ³He has an anisotropic spatial character. The fluid in all the superfluid phases has an intrinsic bending energy that favors the persistence of a particular orientation of the pair wave functions over rather long ranges in the liquid. This property of superfluid ³He has been given the name texture.

At surfaces the wave function must be oriented with the orbital axis normal to the surface (just as the stable angular momentum configuration of a top is the one with the spin perpendicular to the surface). It costs energy for the wave function to change direction in space. However, the textures can be readily bent through a variety of means such as magnetic fields and fluid flow. The characteristic bending length of the texture in the B phase has been found experimentally to

TABLE 8.1 Properties of the Superfluid Phases of Liquid ^{3}He and ^{4}He, of Normal Liquid ^{3}He, and of Superconductors

	Spin Polarization S_z	Orbital Angular Momentum L_z	Total Momentum J	Magnetic Susceptibility	Orientational Dependence of Flow	NMR	Ultrasound (Zero Sound Modes)	Other Sound Modes
Superfluid Phases of ^{3}He								
B	+1 0 −1	+1 0 −1	0	Decreases with decreasing T (spontaneous magnetization of vortices)	Independent of direction of H relative to direction of flow	No shift in transverse resonance in an "ideal container" longitudinal ringing and parallel resonance	$J = 2$ and $J = 4$ distortions, Zeeman splitting of $J = 2$ distortion, pair breaking	Fourth sound (superfluid compression wave) First sound (ordinary sound)

		good quantum number	T dependence	Flow / direction of H	Resonance	Propagation depends	Fourth sound / First sound
(—)	−1	good quantum number	of T as in the normal liquid	rapid along direction of H than perpendicular to it	transverse resonance, longitudinal ringing, and parallel resonance	on orientation of magnetic field, pair breaking, wave function distortions called flapping, and clapping modes	First sound
A₁	+1, −1	Not a good quantum number	Independent of T	Flow more rapid along direction of H than perpendicular to it	Shift in transverse resonance only	Expected to be similar to A phase	Second sound (entropy wave) is also a spin wave
Superconductors	0	0	Decreases with decreasing T	Independent of direction of H	No shift or longitudinal ringing	Pair breaking	None
Superfluid ⁴He	None	—	None	None	No resonance	None	Five other modes
Normal liquid ³He and electrons in normal metals	Uncorrelated spin-½ fermions	None	Independent of T at low temperatures	None	No shift or longitudinal ringing	Zero sound in ³He	Spin waves

be inversely proportional to the field strength and is of the order of millimeters in a field of 20 gauss.

A striking difference between superfluid ^{4}He and superfluid ^{3}He was found in studies of the fluid flow. In both substances the fluid can flow through spaces too small for penetration by the normal-state liquid. In liquid ^{4}He, there are no preferred orientations for such a flow. However, when the flow of superfluid ^{3}He in the A phase was studied in magnetic fields, it was found that the flow was much more rapid along the direction of the field than in the plane perpendicular to the field. The magnetic field aligned the fluid texture to produce an anisotropy in one of its most fundamental properties, the fluid flow. When the same flow experiments were repeated in the B phase, no dependence of the flow on the direction of the magnetic field was observed.

NUCLEAR MAGNETIC RESONANCE IN SUPERFLUID ^{3}HE

The macroscopic quantum nature of ^{3}He has profound effects on its NMR. In liquid ^{3}He, the nuclear magnetic dipoles exhibit a coherent response to external perturbations and are coupled to the orbital character of the macroscopic state. Because of this the frequency of the NMR absorption is shifted as if it were in internal fields of the order of 100 gauss.

Both the A and B phases have a longitudinal resonance not found in NMR in any other physical system except solid ^{3}He. A step change in the magnetic field produces a ringing in the amplitude of the magnetization parallel to the field. The ringing frequency depends on the temperature ratio T/T_c. Similarly, a radio-frequency field polarized along the steady external field can produce a detectable resonant energy absorption at the same ringing frequency.

ULTRASOUND

It has been possible to use high-frequency sound to perform an unusual type of spectroscopy in superfluid ^{3}He. Sound at frequencies much larger than the atomic collision frequency of the liquid, the so-called zero sound, can be propagated with only weak damping and has been studied in all the superfluid phases of ^{3}He and in the normal fluid. Unlike ^{4}He, where only a single zero sound mode exists, a large number of narrow absorption modes have been observed in superfluid ^{3}He because of the complex nature of the order parameter. The excitations associated with the distortion of the order parameter have been studied most extensively in the B phase because its isotropic order parameter makes the experiments easier to interpret. They correspond to the resonant excitation of $J \neq 0$ states from the $J = 0$ ground state. Careful examination of the modes revealed that they were analogues of phenomena first studied for single particles in the early days of atomic physics, the Stark effect and the Zeeman effect. In the case of ^{3}He there is a remarkable difference. These line-splitting phenomena occur because of distortions that affect the wave function that simultaneously describes the behavior of all the atoms in the container at once.

OTHER SOUND MODES

Most of the sound modes that have long been studied in superfluid ^{4}He have also been fruitfully examined in superfluid ^{3}He. Normal, or first, sound, the usual long-wavelength mode of most fluids, has been used to determine the viscosity parameters of the fluid. Second sound, the famous entropy wave in ^{4}He, is also a spin wave in ^{3}He. The $S_z = +1$ polarization of the spin triplet in the A_1 phase was determined through comparing a mechanically induced second-sound pulse with the magnetization change measured in an NMR coil. Fourth sound, a compression wave in a superleak, was used to make the first convincing demonstration of superfluid flow in ^{3}He.

DEFECTS

Sudden discontinuities in the pair-wave functions are called defects. In any real container there must be at least one defect because it is impossible to meet the boundary condition that the wave function be perpendicular to the surface at all points in even the simplest three-dimensional geometry, the sphere. Modern topological methods have been used to classify the types of defects that might appear in the texture fabric.

Sharp walls in which the direction of the orbital axis is reversed by 180 degrees are solitons. Solitons can be created in either the spin or the spatial portion of the superfluid wave function through the application of intense radio-frequency pulses at the NMR resonance. The existence and specific nature of such defects have been investigated through examination of, and the changes in, the resonance modes in the NMR spectrum. These types of defects persist in the liquid almost indefinitely after their creation if neither the fluid temperature nor the magnetic field is changed. They can be erased by removing the magnetic field or by flow of the fluid.

Vortices, which are the only defect known in superfluid ^{4}He, have been studied in both the A and B phases in an apparatus capable of producing a steady rotation of the fluid. The presence of the vortices was demonstrated, and estimates of their density made, again through NMR methods. Recently it has been found that a spontaneous magnetization appears in the vortex cores of fluid in the B phase. This result has strong implications for our understanding of the magnetic field about neutron stars. Modern models of the dense centers of such stars suggest that there also the fermions, neutrons in this case, become paired in a triplet state similar to the B phase of superfluid ^{3}He.

SUPERFLUID FLOW AND HYDRODYNAMICS

In addition to the rotation experiments discussed above, a variety of flow and hydrodynamic measurements have been performed. Fluid flow has been studied in cylindrical, spherical, and parallel plate geometries. The motion of a wire moving through the fluid has also been carefully analyzed. The critical velocities limiting the superfluid flow in both the A and B phases appear to be much smaller than originally expected. The viscosity measured in the fluid at low temperatures is also much smaller than expected. These deviations from theory are probably related to both the long mean free path of the normal fluid excitations and the peculiar kinetic behavior of these excitations at boundaries.

One useful product of the flow studies has been the determination of the superfluid mass fraction in ^{3}He. The results are important for understanding other hydrodynamic experiments. The superfluid mass fraction can be found from determinations of the amount of fluid that is not changed through viscous contact with chamber surfaces.

Novel Quantum Fluids

There is a variety of gases and fluids that can be cooled to low enough temperatures for quantum-mechanical effects to become important in understanding their physical behavior. Examples that have been studied for quite some time include the bulk behavior of liquid ^{3}He, liquid ^{4}He, liquid mixtures of the two helium isotopes, and the conduction electrons in metals. The temperature at which quantum statistics becomes important is that at which the thermal de Broglie wavelength becomes comparable to the interparticle spacing. In the case of fermions, ^{3}He, and electrons, this temperature is the Fermi temperature, and in the case of bosons, for example ^{4}He, the temperature is roughly the Bose condensation temperature. Statistical mechanics predicts a macroscopic occupation of the ground state for Bose particles cooled below the Bose condensation temperature. In liquid ^{4}He this temperature is calculated to be 3 K, and it seems likely that the superfluid transition of bulk ^{4}He at 2.2 K is related to this result.

In the latter part of the 1970s attention was turned to these fluids under new, extreme physical conditions. The Fermi systems are being investigated under conditions of large polarization where only one spin population is present. The Bose systems are being investigated in unusual geometries and under conditions of increasing dilution of the distance between the particles.

The newest materials in this class that have come under investigation are the gases of atomic hydrogen and deuterium, which can be stabilized against the formation of H_2 and D_2 through polarization of the atomic electrons in large magnetic fields at low temperatures.

MIXTURES OF ^{3}HE IN ^{4}HE

At low temperatures ^{3}He is soluble in ^{4}He in concentrations up to 6 percent for fluid with no external pressure, and up to 10 percent for fluid under pressures greater than 10 bar. The gas fermions in superfluid ^{4}He have weak interactions. The ^{4}He acts as an ether medium supporting a dilute gas that behaves almost like an ideal Fermi gas. The properties of such mixtures in low magnetic fields were studied extensively in the late 1960s and early 1970s to measure the small deviations from ideal gas behavior and to develop the dilution refrigerator (this is discussed later in this chapter in the section on Low-Temperature Technology).

SPIN-POLARIZED HYDROGEN AND DEUTERIUM

Atomic hydrogen and deuterium can be stabilized against the formation of molecules if their electronic spins are polarized. The spin-triplet potential has no bound states. Hence, if all the atoms could be forced to interact only through the triplet potential large densities of the material could be collected, and because of the weak interaction between the atoms they would remain as a gas to absolute zero temperature. The hydrogen atom has an even number of fundamental particles, one electron and one proton. It thus is a boson and should display a statistical condensation like that in ^{4}He. Deuterium is a fermion, and it should have properties similar to the weakly interacting Fermi gases discussed in other sections of this report. The search for

Bose condensation in atomic hydrogen has been a topic that has attracted a great deal of interest in the early 1980s. It offers an entirely new example of a superfluid that can be compared with ^{4}He. Hydrogen would be an especially important example because the interparticle coupling is so weak that virtually all of the substance properties can be calculated from simple principles. In addition, hydrogen has a nuclear magnetic moment, whereas ^{4}He does not. A large class of collective spin phenomena, similar to those in superfluid ^{3}He, are expected to be found.

Research in this area is in its infancy. The experimental problems are formidable, but significant progress has been made. Atomic hydrogen densities of 10^{18} atoms per cm^3 have been achieved at a temperature of 0.5 K. The Bose condensation temperature for such a density is ~ 17 mK. At a temperature of 100 mK a density of 10^{19} atoms is required (T_c varies as the 2/3 power of the density) for Bose condensation. It is not obvious that the problem can be solved by improvements in the cooling techniques. The hydrogen in the gas phase is in pressure equilibrium with atoms bound to the container surface. The dominant processes for atomic recombination take place on the surface. The surface recombination rate increases rapidly with the density of surface atoms. The lowest binding energy of hydrogen bound on any surface is that on helium films. However, even using surfaces preplated with liquid helium, gas densities of 10^{17} atoms per cm^3 would be expected to saturate the surface completely at temperatures of a few millikelvins. On the other hand, the time constants for the decay of the collected hydrogen can be quite long—several hours. Several research groups have succeeded in performing transient experiments in which the hydrogen density is increased to $\sim 3 \times 10^{18}$/cm^3 by compressing a bubble of the gas in liquid helium.

LIQUID ^{4}HE IN UNUSUAL GEOMETRIES

During the past decade, the superfluid transition in liquid ^{4}He has been extensively studied in thin films on a variety of substrates. Films deposited on smooth and flat surfaces provided the first model for measuring the way in which the order of a two-dimensional system is disrupted by thermally activated defects. In helium films, the order in the low-temperature phase appears in the fluid momentum. The inertial response of the mobile atoms in the film is correlated. As the temperature is raised, the order is disrupted by thermally activated vortex pairs. When the vortex density becomes sufficiently large, the long-range order in the fluid motion becomes completely disrupted. The film then responds in the same way as a normal fluid with viscous damping of the substrate-induced motion. Analogous phenomena have been studied subsequently in a variety of other physical systems. Some examples have been the melting transition in two-dimensional crystals, the disorder transition in some liquid crystals, and the normal to superconducting transition in thin metal films. Superfluid ^{4}He provided the ideal test substance for such studies because of its relative cleanliness. There are few defects other than those specifically under investigation.

In another extension of superfluid ^{4}He to new regimes, the transition of thin films deposited in a porous three-dimensional glass network has been studied at very low temperatures. In this system, T_c decreases as the amount of fluid deposited on the substrate decreases. For thick films with a high transition temperature, the character of the superfluid mass change below T_c appears much like that in the bulk fluid where critical fluctuations dominate the behavior. In experiments using the lowest density studied to date, the superfluid interparticle spacing is more than an order of magnitude greater than the hard-core atomic diameter. At these low densities, the character of the transitions appears to be changing. The superfluid mass seems to be approaching a linear variation in $T_c - T$, the behavior expected for the dilute Bose gas without critical fluctuations. The work to date extends down to temperatures near 5 mK. Our understanding of the bulk superfluid transition is incomplete. The interactions among the atoms are so strong that no one has succeeded in making a microscopic theory that can account for the superfluid state. We may gain valuable insight about how to solve the problem as techniques are developed for studying the transition at even lower temperatures. The evolution of the diluted gas toward complete behavior like that of the weakly interacting Bose gas of textbooks would be an important advance for this fundamental problem.

ELECTRONS ON HELIUM SURFACES

An electron in the vacuum above the planar surface of a dielectric medium gives rise to an image potential in whose discrete eigenstates it is bound to the surface. A static electric field directed normal to the surface can further clamp the electrons to the surface, as well as vary their areal density on the surface. By these methods electrons have been trapped above the free surface of liquid ^{4}He and between phase-separated liquid mixtures of ^{3}He and ^{4}He. Since their motion normal to the surface is quantized the electrons form a two-dimensional model system on the surface that is appealing for its simplicity. When the electron density on the free surface is sufficiently large, and the temperature is low enough, the electrons form a two-dimensional crystal called the Wigner lattice. This crystallization of the electrons was detected 5 years ago through studies of the vibrational modes of the resulting electron crystal on the helium surface, as well as by the change in the mobility of the electrons along the surface on crystallization. The system of electrons on surfaces of helium is an ideal model of the defect-free two-dimensional conductor. Current research is addressed toward general questions that are difficult to study in films of metals and semiconductors because of defects and impurities.

SUPERCONDUCTIVITY

Superconductivity is a phenomenon of great intricacy, diversity, and elegance. It is one of the most interesting and intellectually challenging subdisciplines of physics and has led as well to remarkable and essential applications in mankind's most ambitious technologies. Intellectual activity in the science of superconductivity remains vigorous today. At the same time there is explosive growth in the technical and industrial application of superconductivity. Large-scale applications include the highest-energy-density electric power systems just now being integrated into the electric utilities, an experimental magnetically levitated train (the world's fastest), magnets used in NMR systems for noninvasive inner-body medical diagnostics imaging (Figure 8.2), the world's largest electromagnets (being used for magnetic-confinement thermonuclear fusion experiments), and bending and focusing magnets for the world's most energetic particle accelerator. Small-scale electronic-type applications include the fastest-switching and least-power-consuming electronic devices, the most sensitive and lowest-noise magnetic and electromagnetic sensors, and the most accurate voltage standard, to name but a few. At present the worldwide economic impact of superconductivity is estimated to be a few hundred million dollars annually. Impacts as great as billions of dollars by 1990 and tens of billions by the year 2000 have been estimated.

These advances in superconductivity have not been easily won. Subsequent to its discovery in 1911, nearly 50 years of intensive experimentation and theoretical development passed before a microscopic theoretical understanding of superconductivity was achieved through the Bardeen-Cooper-Schrieffer (BCS) theory. In fast succession thereafter, in the early 1960s, two significant scientific advances

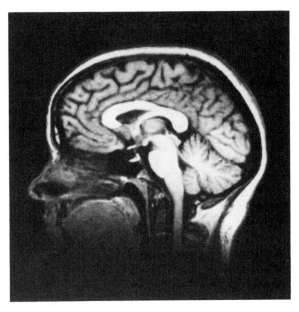

FIGURE 8.2 Midline sagittal view of normal human head. In the nuclear magnetic resonance technique used to construct this image the subject was positioned in the magnetic field of a 1.5-T superconducting magnet. (Reproduced with permission from General Electric Company.)

occurred, which not only paved the way for the rapid growth in applications of superconductivity during the 1970s and 1980s but also set much of the agenda for the scientific endeavors of the current era. First was the discovery that certain of the superconductors now known as type II superconductors can, without power dissipation, support extremely high electric current densities ($\sim 10^5$ A/cm^2) despite the presence of very high magnetic fields (~ 10 T). This provided the basis for large-scale utilization of type II materials in superconducting magnets in technological applications. Theoretical understanding of many of the remarkable features of type II materials was soon achieved in terms of extreme parametric regimes of existing phenomenological theories of Ginzburg, Landau, and Abrikosov, which had been linked to BCS theory by Gor'kov.

The second highly significant scientific advance was the prediction and subsequent experimental confirmation of the Josephson effects, viz., that supercurrents (dissipationless currents) can tunnel between superconductors despite their being separated by thin, normally insulating barriers, that the maximum dissipationless current is a sensitive periodic function of magnetic flux (periodic in the flux quantum), and that when the critical Josephson current is exceeded an ac signal is generated with a frequency linearly proportional to the junction voltage.

The uniqueness of superconducting macroscopic quantum phenomena, the beauty of the successful theoretical descriptions, the proven applications, and the high potential for still wider applications all combine to make superconductivity an attractive and challenging area for further research and development.

In the past decade the scope of superconductivity research has broadened significantly, and the pace of discovery has continued unchecked as new directions and focal points have emerged. For example: (1) Progress was achieved in understanding the role of thermodynamic fluctuations in determining the nature of the superconducting transition in both bulk and restricted geometries. (2) Superconductivity has continued to be an important theme of theoretical and experimental efforts in many-body phenomena and has been used as a means of testing new models and descriptions of physical systems. (3) Wide-ranging investigations have clarified the consequences of dynamic effects and nonequilibrium in superconducting systems. (4) Superconducting systems have been used in examinations of fundamental questions in statistical physics and quantum mechanics, including quantum noise, special types of two-dimensional phase transitions, and, most recently, macroscopic quantum tunneling and chaotic behavior. (5) Superconductivity has been discovered and studied in depth in novel and exotic materials such as quasi-one-dimensional polymeric and organic materials, quasi-two-dimensional intercalated-layer compounds, artificially layered or compositionally modulated materials, inhomogeneous materials, metastable materials, and so-called reentrant ferromagnetic superconductors. These experiments yielded answers to fundamental questions concerning the normal electronic structures of these exotic materials. (6) Greater understanding and improved performance have been achieved for high-magnetic-field superconducting materials of technological interest for large-scale applications, and at the same time the search for new still-higher-performance, high-magnetic-field, high-transition-temperature materials has been vigorously pursued. (7) Combined physics, materials, and device efforts on the many aspects of the Josephson effects have led to greatly expanded understanding of these phenomena and to new and improved magnetic and electromagnetic sensors, the most sensitive known; to the fastest known signal-processing circuits (sampling circuits, convolvers, and analogue/digital converters); and to major progress toward a very-high-performance, Josephson-effect-based computer, potentially the highest-speed, most-compact, and most-power-efficient approach known.

The above abbreviated list is necessarily only representative of the much broader totality of superconductivity research. Even with that restriction a comprehensive account of the major developments in each of these areas cannot be presented here, and so what follows are brief descriptions and highlights selected from the above areas.

Nonequilibrium Superconductivity

The study of nonequilibrium superconductivity is the study of the response of superconductors to externally applied perturbations. Because superconductors have much longer relaxation times than normal metals, they can be driven out of equilibrium more readily than normal metals. For example, the electronic and ionic (phonon) effective temperatures can be decoupled in superconductors. This observation, together with the seminal discovery that the chemical potentials of the excitations and condensate in a superconductor could be made significantly different by the injection of charged particles, and could readily be measured, opened up a fertile and exciting area of many-body, condensed-matter research. The rich variety of studies in this field has included the microwave radiation enhancement of the superconducting energy gap, the explanation of excitation-to-condensate conversion processes, the elucidation of unusual behavior in superconducting microstructure and contacts, and the as-yet-unexplained anomalously large thermoelectric generation of flux in a superconducting bimetallic loop. More recent studies focus on regimes far from equilibrium such as the transition to the normal state triggered by supercritical currents. Although much progress in this area has been made, many research opportunities remain, particularly in the area of extreme nonequilibrium conditions.

Novel Superconducting Materials

Searches for superconductivity in a variety of novel or exotic materials have been richly rewarded in the past decade. Some quasi-one-dimensional systems, such as $TaSe_3$, $NbSe_3$, and polymerized sulfur nitride, and quasi-two-dimensional conductors, such as intercalated dichalcogenides and artificially layered structures, have been found to superconduct. Superconductivity has been found in organic systems, the first of which was $(TMTSF)_2PF_6$.

New metallic compounds have been found to have unusual superconducting properties. For example the Chevrel compounds have high transition temperatures and record critical fields from 50 to 70 T. Superconductivity has been observed in low-electron-density compounds such as $BaPb_{1-x}B_xO_3$. Finally, heavy-fermion superconductors with enormous electronic specific heats have been found. The first example was $CeSi_2Cu_2$, and its discovery was followed by the discovery of three more heavy-fermion superconductors, UBe_{13}, $CeCu_6$, and UPt_3. These systems are particularly exciting because they may be the first examples of superconductors in which the electrons are paired with parallel spins.

Renewed interest is developing in amorphous and metastable systems such as the record transition temperature A15 materials. Still more unusual phenomena await diligent materials synthesists.

Magnetic Superconductors

In 1958 it was shown that as little as 1 percent of magnetic impurities can destroy superconductivity in the host metal. Beginning in the mid-1970s, the interplay between superconductivity and magnetism has been re-explored. Unexpected results were found in a family of ternary rare-earth compounds typified by $ErRh_4B_4$. In these compounds the superconductivity is associated with the transition-metal electrons that are confined in clusters and are therefore relatively isolated from the magnetic rare-earth ions. This sets up a competition between superconductivity and magnetism that reveals itself in unusual behavior. For example, $ErRh_4B_4$ becomes a superconductor near 9 K. As it is cooled further, the rare-earth ions begin to order magnetically until the superconductivity is destroyed near the Curie temperature just below 1 K. Recent small-angle neutron-scattering experiments suggest that some of these compounds exhibit a new phase of matter in which superconductivity coexists with magnetic order in periodic structures with a wavelength of about 200 Å with superconductivity surviving.

High-Transition-Temperature, High-Magnetic-Field Materials

Efforts to increase the high-magnetic-field performance of superconducting devices have gone in two directions. One is the improvement of existing materials, and the other is the search for new materials.

The past decade has seen striking progress in the technical use of existing supermagnet materials. Part of the progress is due to improved understanding of flux jumping and related thermal effects. This has led to the development of multifilamentary cables optimized for maximum performance in specific applications.

Progress has also been made through painstaking improvement in production processes. Even so, studies of optimized short samples show that much improvement in commercial materials can still be made.

The quest for new technologically tractable materials has proved more difficult. Although Nb_3Sn with its superior superconducting properties has been fully stabilized, the more ductile Nb-Ti alloys, currently produced in the thousands of tons, remain the workhorse of high-field superconductors (Figure 8.3). Nevertheless, as noted earlier, significant progress in the understanding of the conditions for the occurrence of superconductivity have been made. Inevitably, this must contribute to the development of truly superior metals.

FIGURE 8.3   High-performance multifilament superconducting conductor used in the Mirror Fusion Test Facility superconducting magnets at Lawrence Livermore National Laboratory. Nb-Ti alloy superconducting filaments (dark regions seen end on in center square) are embedded in normal metal matrix. (Courtesy of Oxford Airco.)

The Josephson Effects

The Josephson effects are among the most beautiful and novel manifestations of the macroscopic quantum nature of superconductivity. Their study and their technological application have continued to be exciting. Central to rapid progress in this area during the past decade has been the highly successful development of techniques for fabricating micrometer- and submicrometer-size junction structures of high quality, uniformity, and reliability. This has made possible the application of nearly ideal Josephson structures to a number of exciting scientific and technological endeavors.

One scientific focus has been the investigation of dynamic Josephson effects in thin-film microbridges. Another has been the study of the quantum limits of noise. The most sensitive magnetic-field detectors, the so-called superconducting quantum interference devices (SQUIDs), have been fabricated with an energy resolution only slightly greater than the uncertainty principle limit h (Figure 8.4). In addition to being direct objects of study, these sensors are the essential components in a wide variety of other low-temperature quantum-noise studies.

Josephson devices are being used to address a variety of fundamental questions in condensed-matter physics. The investigation of macroscopic quantum tunneling is a current, active example. The purpose of this work is to test efforts aimed at including dissipation (friction) in quantum-mechanical descriptions of macroscopic physical systems. This work is just getting under way and appears to have a bright future. The use of large, two-dimensional arrays of Josephson junctions to examine concepts of two-dimensional phase transitions is but another example of how superconductivity serves as a novel means of testing general theoretical hypotheses. The study of chaotic behavior in Josephson junctions is another recent example.

QUANTUM CRYSTALS

Quantum crystals are solids in which the atoms have a large-amplitude zero-point motion. The most important quantum crystals are helium, hydrogen, and deuterium, the same elements found in quantum fluids. In the case of hydrogen and deuterium, the atoms are in a molecular form, and interest in these crystals centers mostly around the fact that they are the simplest of the molecular crystals. The molecules solidify in several angular momentum states, and the low-temperature properties including the crystalline structure depend on

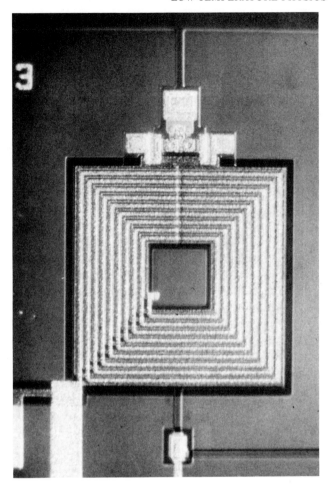

FIGURE 8.4   Ultralow-noise Josephson analog superconducting quantum interference device (SQUID) with spiral input coil. Devices such as this are utilized for extremely high-sensitivity magnetic-field measurements. (Courtesy of IBM Corporation.)

the fraction of the molecules that have decayed to the lowest angular momentum state. The molecules in solid hydrogen and deuterium interact by means of a quadrupolar interaction. Crystalline fields in the solid couple with the interatomic quadrupolar field of the molecule.

The quantum motion is much larger in helium than in solid hydrogen because of the weaker binding potential. A major consequence of the quantum motion in solid ^{3}He appears in its magnetic properties. The overlap between the wave functions of neighboring atoms leads to an atomic exchange. The exchange energy is of the order of 1 mK for crystals with the largest molar volume, at the melting pressure. The size of the exchange energy decreases rapidly with decreasing molar volume V of the solid, varying approximately as V^{18}. The large exchange energy in solid ^{3}He produces magnetic order in the crystal at 1 mK. The transition occurs at the highest temperature of any nuclear magnetic transition. In copper, for example, the nuclear ordering transition occurs at 60 nK. The details of the solid ^{3}He magnetic transition are still under intensive investigation. There are quite a number of surprising features to the transition. It is first order in nature, and the entropy drops discontinuously by a factor of 0.44 R ln 2 at T_c. The solid expands in volume by 1×10^4 at the transition, and the magnetic susceptibility drops by more than a factor of 2 at T_c. The microscopic nature of the sublattice orientation was determined through an elegant set of nuclear resonance measurements in which the antiferromagnetic spin-wave modes were studied. The crystal has a body-centered cubic lattice, and the sublattice structure is one in which the direction of the spins on successive planes alternates every two planes. The sublattice planes are parallel to the cubic faces of the crystal. The up-down-down structure has been abbreviated u2d2. This low-field phase is unstable in magnetic fields greater than 0.45 T. In fields greater than 0.45 T another magnetic phase appears. Little is known about the microscopic nature of the high-field phase. It is likely to be similar to the spin-flopped phase often found in antiferromagnets.

The most successful theory of the magnetic properties of solid ^{3}He suggests that the transition occurs as the result of the competition between three and four particle rings of exchange in the bcc crystal. The odd number of interchanges favors a ferromagnetic order, and the even number favors an antiferromagnetic order. Using plausible values for the exchange rates, a two-parameter theory seems to give reasonably accurate descriptions of the observed phenomena.

There have been recent investigations into the nature of the interface between liquid and solid helium, to determine whether the surface has facets at low temperatures. In all other crystals there is a roughening transition at which the low-temperature state, with flat faces related to the crystal structure, is disrupted by thermally activated defects so that the facets disappear. It had been speculated that such a transition would not exist in solid helium because the zero-point motion would keep the surface rough even at zero temperature. However, it was found that facets do, in fact, appear in crystals of solid ^{4}He. Experiments have not yet been performed on solid ^{3}He, which has a larger zero-point motion.

LOW-TEMPERATURE TECHNOLOGY

One of the most important traditions of this research field has been that of extending the experimental working regime of physical measurements to lower temperatures. The historical progress of the field is illustrated in Figure 8.5, where the minimum equilibrium temperature achieved after each new technological advance is plotted versus the year of the advance. The minimum temperature has decreased by a factor of 10 approximately every 15 years, since the first liquefaction of air a little over 100 years ago. Despite the apparently continuous nature of the progress suggested by the graph, the new advances have always come after long periods of consolidation of the most recently developed methods. As is the case with many other technologies, the cryogenic advances have been transferred to related areas of scientific research. This has typically happened 10 years after the workers in the low-temperature physics community have consolidated the experimental method. The use of apparatus that requires working temperatures of 4 K is common today. The present frontier of the field is in the region of 10 μK.

The cooling method in modern apparatuses working at the lowest

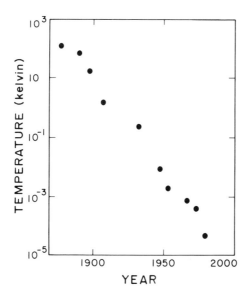

FIGURE 8.5   Graph representing progress in cryogenic technology. Each point in this semilogarithmic graph represents a major advance in technology.

temperatures relies on two technologies that have been refined in the 1970s, the dilution refrigerator and superconducting magnets. The dilution refrigerator is an apparatus that takes advantage of the phase separation in mixtures of ^{3}He in ^{4}He. ^{3}He is driven from a ^{3}He-rich phase into the dilute phase by a concentration gradient in the dilute phase. The thermodynamic process is similar to the evaporation of atoms from a liquid to a gaseous phase. There is a large entropy increase when the ^{3}He atom passes from the ^{3}He-rich phase into the dilute phase. After the ^{3}He atom passes through a dilute column of liquid mixture, it is evaporated and collected by pumps at room temperature. The ^{3}He gas is then recycled, and after heat exchange with the dilute column returns to the ^{3}He-rich phase at low temperatures. The process can operate continuously, and temperatures as low as 2 mK have been maintained with this cycle. Equipment of this type for experiments down to 5 mK can now be purchased from several different commercial manufacturers. The combination of a constant low-temperature thermal sink at 5 mK and modern superconducting magnets capable of producing fields of 15 T make it possible to achieve large values of the ratio B/T, the significant variable for achieving large polarizations in any paramagnetic system.

To obtain even lower temperatures, magnetic cooling cycles are employed. The dilution refrigerator is used to precool a paramagnetic refrigerant placed in a large magnetic field. The refrigerant is thermally isolated after it has been polarized. The next stage of cooling is achieved by reducing the magnetic field. Nuclear moments are used for the refrigerant material. A significant simplification of the magnetic cooling technology was accomplished by the introduction of a new magnetic material, $PrNi_5$. In this material, there is a hyperfine interaction between the electrons and the nucleus of the Pr atoms. The hyperfine field acts as an amplifier for the external applied field. The local field at the site of the nuclei is enhanced by more than a factor of 20. Complete polarization of the Pr nuclei can be achieved with a modest dilution refrigerator, operating down to 10 mK, and an 8-T magnet constructed with NbTi wire, the least expensive magnet wire. Unfortunately, the hyperfine interaction that assists in producing the large polarization also limits the minimum temperature of the material. The minimum temperature is 0.4 mK, a limit imposed by magnetic order in the metal.

The lowest temperatures have been obtained by using two cascaded stages of nuclear magnetic cooling. Typically $PrNi_5$ has been used as the first magnetic stage, to remove the heat of magnetization from copper. After the copper has been demagnetized, stable lattice temperatures as low as 20 μK have been measured. The spin temperature of the copper goes much lower. A nuclear ordering transition has been measured in the copper nuclei at a temperature of 60 nK. It is not clear what limits this technology. Only a few materials have been tested. Metals with good electrical conductivity must be used to have reasonable equilibrium times in the lowest temperature stages. The ultimate lattice temperature is governed by the balance between the metal conductivity and the heat leak from external sources. In the most successful apparatuses, it has been estimated that the heat leak from cosmic radiation might be playing a significant role in determining the minimum temperature.

RESEARCH OPPORTUNITIES IN LOW-TEMPERATURE PHYSICS

Superfluid ^{3}He has been studied for one decade, and a great number of interesting questions remain to be answered. The analogues of some of the most important superconductivity experiments have not yet been repeated. For example, the Josephson effect should be observable in ^{3}He. Persistent currents have never been created in superfluid ^{3}He, and the question of the quantization of fluid circulation in the various phases has not been tested. Most of the texture phenomena suggested by theory remain to be investigated. Size effects and the limits on the superfluidity imposed by the dimensionality of the fluid container have not been studied.

The Fermi temperature of pure liquid ^{3}He is roughly 1 K. It is not practical to produce a highly polarized specimen through the brute-force application of a large magnetic field (at any temperature), because the 1300-T field required is several orders of magnitude too large for present technology. Instead, several groups are developing transient methods for polarizing liquid ^{3}He. Two methods look promising. In one case, the ^{3}He starts off in the solid phase at 34 bar, where it is polarized by cooling at low temperatures in a large magnetic field. It is then converted to the liquid phase by expansion of the chamber volume. In the second method, the ^{3}He is polarized by optical pumping of the atoms in the gaseous phase at room temperature. The gas is then cooled to the condensation temperature to produce a polarized liquid. In both cases, the lifetime of the polarized state is limited by spin-relaxation processes at surfaces or interfaces. Under some circumstances these times can be long enough that interesting measurements are possible. When there is only one spin population present, the fluid properties are likely to be different from those of unpolarized liquid

^{3}He because of changes in the interactions between particles. When the nuclear spins are parallel, the Pauli principle requires them to have a larger separation than atoms when the spins are antiparallel. Thus, the density of the fluid will change slightly. The average interaction between the ^{3}He atoms should be different from those in unpolarized liquid ^{3}He so that quantities like the heat capacity and the magnetic susceptibility are expected to change. From a theoretical standpoint, the properties of the polarized liquid are expected to be easier to calculate because the interatomic potential is simpler.

There are two areas of especially interesting research with mixtures of ^{3}He in ^{4}He that are likely to be significant in the coming decade. One is the search for a pairing transition between ^{3}He atoms as the fluid is cooled to lower temperatures. The second is the study of the transport properties of the fluid in large magnetic fields.

There are no reliable estimates for the pairing transition temperature in the dilute mixtures; nor is it known whether the pairing will be a triplet state like that in pure liquid ^{3}He or a singlet state similar to that in superconductors. In pure ^{3}He and in superconductors T_c is $\sim 10^{-3} T_f$, where T_f is the Fermi temperature. With each advance in cryogenic technology the dilute mixtures have been re-examined to see whether there are sudden changes in the magnetic susceptibility or heat capacity that would mark the onset of the pairing transition. By 1983 the experiments had been extended down to a little over 200 μK for solutions with a Fermi temperature of 100 mK. No pairing transition has yet been observed.

The dilute mixtures are technically easier to polarize than pure ^{3}He because the Fermi temperature can be adjusted to be small enough to match the polarization energy available with practical magnets. For a mixture with a ^{3}He fraction of 10^{-4}, the Fermi temperature is 5 mK. When such a solution is cooled to a few millikelvins the nuclear moments can be almost completely polarized in a magnetic field of 8 T, a field that is relatively easy to obtain. To date, there have been few investigations of this system, but the fluid should have some remarkable properties. The scattering cross section for polarized ^{3}He atoms in solution is much weaker than that of the unpolarized atoms. Thus, the mean free path between particle collisions is expected to be more than a factor of 10 longer than that in the unpolarized liquid. The most interesting quantities to measure are the transport coefficients: the thermal conductivity should be larger than that of good metals; the viscosity should approach that of liquids about to form a glass; and the diffusion coefficient should be larger than that measured in any other liquid. Many fruitful studies are likely to come from this system in the near future.

Even if the Bose condensation of spin-polarized hydrogen is not achieved, there are likely to be many useful by-products of the research that should have an impact on technology. Studies of wall relaxation phenomena have already led to a specific design for a better frequency standard. Other technically significant developments are likely to be an improvement of the techniques for storing excited atomic populations for work with lasers and the demonstration of a particularly clean system for studies of chemical reaction kinetics in hydrogen.

The future of superconductivity in the next 10 years seems to be readily apparent from its past 10 years. Even without the unpredictable discovery of a material with a much higher T_c, of an alternative pairing mechanism other than the electron-phonon interaction, or of new phenomena with the impact of the Josephson effects, the field is likely to continue to prosper along the lines of the recent past. It is reasonable to predict that new and unusual superconducting materials will continue to be discovered and avidly studied. Future improvements in the theory of dynamic phenomena in superconductors seem likely—so are improvements in device performance and in high field materials. There also seems as yet to be no end to the use of superconductivity as the test vehicle of basic theoretical advances in condensed-matter science.

A question of current interest in studies of molecular solids is that of how order is achieved in the orientation of $J = 1$ molecules. There are speculations that a glassy state of order exists in which there is short-range orientational order but no long-range order. Nuclear resonance experiments have been performed that support both sides of the argument. This is an issue that is likely to be resolved in the near future.

9

Liquid-State Physics

CLASSICAL LIQUIDS

Whereas a crystalline solid is invariant against displacement through a lattice constant along each of the three principal coordinate axes, a liquid is invariant against an arbitrary displacement in space. A liquid also differs from a crystalline solid in its orientational symmetry. In a crystal the bonds or lines joining nearest neighbor atoms are oriented along specific directions in space. In a liquid, however, the lines joining pairs of nearest neighbor atoms will point with equal probability in all directions of space. There also exist in nature various liquid-crystal phases, which exhibit a broken orientational symmetry, like a crystal, but possess the translational invariance of a liquid. In this chapter we survey recent advances in our understanding of classical liquids and of liquid crystals, and point to areas of liquid-state physics in which progress can be expected in the next few years.

Introduction

In terms of everyday experience, liquids are certainly as common as solids. The study of liquids has a renowned classical tradition centering largely on the great disciplines of hydrodynamics and hydraulics. However, there is a more atomistic aspect of the study of liquids that has paralleled some of the developments in the statistical mechanics of

190

solids, though it has progressed more slowly. Research is focused on the microscopic description of liquids, which in the last decade has seen noticeable advances both in theory and in experiment. In particular, the whole notion of experiment has broadened to include certain Monte Carlo and molecular-dynamical computer simulations noted below.

The intent of the microscopic view of fluids is to try to understand the static and dynamic properties of fluids, typically in the classical regime (where quantum effects are unimportant), starting from the basic principles of classical statistical mechanics and a knowledge of the fundamental interactions in the system. These interactions represent the basic forces between the atoms or molecules of the liquid and change according to the type of system being discussed (e.g., liquid argon, liquid metals, and molten salts). On the scale of thermal energies the interactions in the above examples are strong, and even simple liquids of monatomic molecules, which have spherically symmetric interactions, are highly correlated systems.

To treat this classical many-body problem, the common starting point is to assume the atoms interact by means of pair forces alone. This is only an approximation (though often a good one) because it is known that the influence of one atom on a second is often modified by the presence of a third. Moreover, full details of the exact pair interactions between most real molecules are not yet known precisely. It is partly for this reason that computer experiments in the past decade have become so valuable a source of information. With these techniques it is now possible to simulate the experimental properties of model pair-potential fluids that are immediately pertinent to the microscopic theories that attempt to explain them. In such hypothetical fluids, only pair interactions are considered and the forces between pairs of particles are unambiguously defined. A small class of representative models (no one of which is intended to mimic any particular fluid exactly) have been studied exhaustively enough to yield reliable benchmark results. With Monte Carlo and molecular-dynamics techniques it has been possible to get accurate data both on the structure of these model fluids and on the major functions that describe their thermodynamic properties.

Static Properties

Liquids, by their very nature, are disordered systems whose physical attributes must be described in statistical terms. More particularly, the structural properties of the liquid are represented in terms of distribution functions that give the probabilities of finding given numbers of atoms (one, two, three, . . .) at certain locations. The most prominent of these functions is the pair-distribution function (see Figure 9.1), which gives the probability of finding an atom at a distance r from a given atom. In real fluids, the pair-distribution function is actually determined by the scattering of x rays or neutrons but is also, however, directly obtainable from simulation methods. One of the main tasks of the theory of classical fluids is to determine these distribution functions starting only with the interactions between particles (generally the pair potentials), the mean density of particles, and the temperature and to deduce the thermodynamic properties of the corresponding fluids from them.

Over the past decade a whole range of methods for finding the pair-distribution function and associated thermodynamics has reached maturity. No longer do workers in the field seek one unique way of predicting liquid properties; instead there is a hierarchy of techniques to choose from in which increasing quantitative accuracy can be had for the price of decreasing analytic simplicity and increasing computational labor. These techniques include thermodynamic perturbation theory and its variants, as well as the use of integral equations for finding approximate pair-distribution functions. The integral equations range from those that can be solved in terms of closed-form expressions (the so-called mean spherical approximation and generalizations thereof) to somewhat more complex equations that must be handled numerically but often yield approximations of even higher accuracy

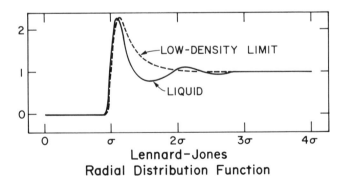

Lennard-Jones
Radial Distribution Function

FIGURE 9.1   Pair-distribution functions for triple-point and low-density fluids. The density of particles, relative to the mean density, is plotted as a function of the particle separation. The distance σ corresponds to the collision diameter for the 6-12 Lennard-Jones pair potential. (Courtesy of W. G. Hoover.)

(e.g., the exponential and renormalized hypernetted chain approximations). Both the computer simulation and the theoretical methods, first applied to simple classical models of monatomic fluids and idealized models of ionic and polar fluids, are now being extended to cope with the presence of intrinsic n-particle forces for $n \geq 3$ as well as with the related but distinct problem of computing n-particle distribution functions for $n \geq 3$ for model pair-potential fluids. Perhaps even more important, over the past 5 years enormous progress has been made on a number of fundamental extensions of the above work. To give some examples: (i) The treatment of nonsimple fluids consisting of polyatomic molecules has yielded to both computer-simulation and integral-equation techniques (often applied to the key probability of simultaneously locating two atoms on different molecules). (ii) Variants of the theories that we have discussed above are also being applied with success to colloidal suspensions and other liquids containing macromolecular particles. (iii) Analytically viable path-integral approaches have been developed to deal with intrinsic quantum effects in the liquid state (e.g., the polarizability of liquids), and, at the same time, powerful computer-simulation methods have been used to solve the Schrödinger equation exactly for many-particle systems under various liquid-state conditions. (iv) The effects of the liquid-state environment on chemical reactions and on conformation changes have begun to be studied in depth using statistical-mechanical models and formulations. (v) Several other technologically important areas of liquid research are also rapidly beginning to reach maturity. The formal theory of inhomogenous fluids was already well developed several decades ago, but the surfaces of liquids and the boundary regions of liquids in contact with solids, which give rise to the wetting problem, are only now being studied intensely, with the promise of reliable predictions for the first time. (vi) For many years, observed liquid-mixture phase diagrams included types that sometimes eluded theoretical realization with Hamiltonian models, even for two-component mixtures. The binary-mixture types all appear to be reproducible theoretically now, although full understanding in this area is far from complete. Mixtures that become unstable and separate under increase in temperature are especially challenging in this connection.

Dynamical Properties of Classical Liquids

The determination of properties associated with molecular motion in condensed phases consists of three approaches: (1) direct experimental measurement of spectral lineshapes, transport coefficients, and relaxation times; (2) analytical or simple model-based theory; and (3) computer simulation of realistic models for fluids.

The experimental techniques used to study fluid dynamics can be divided into two categories: those that probe single-particle dynamics and those that probe collective (many-body) motions. Within each category there are numerous techniques that often provide complementary information and that provide probes of dynamics over a wide range of time (or frequency) and wavelength scales. Nuclear magnetic resonance (NMR), electron spin resonance (ESR), infrared and Raman spectroscopy, and a host of relatively new nonlinear optical techniques fall within the first category. Using these methods one can obtain relaxation times associated with phenomena such as molecular rotation, vibrational relaxation and dephasing, and intramolecular rearrangements. To elucidate the physics that determines these time scales, one makes measurements over a range of physical conditions, for instance over a range of temperatures. Recent experiments employing pressure or density as an external variable have had particular impact. For example, ESR studies have shown that simple free volume corrections to the Debye-Stokes relation for rotational diffusion times, which seem to work well in describing the temperature dependence, may not be valid over a wide range of pressures. Studies of vibrational lineshapes as functions of temperature and density have provided information that has stimulated the development of the first comprehensive theory grounded in a fundamental treatment of intermolecular forces and time scales. NMR studies of the pressure or density dependence of intramolecular rearrangement rates in small alkanes have provided the first experimental evidence that such rates decrease at low densities, a result in marked contrast to the predictions of transition rate theory but in accord with recent theoretical predictions based on the premise that reactions in fluids are friction controlled and require energy dissipation. The advent of picosecond and subpicosecond laser techniques has also led to advances in the understanding of dynamics in fluids, allowing fast processes to be studied directly in the time domain. Such studies have allowed at least partial separation of fast and slow, or homogeneous and inhomogeneous, contributions to vibrational relaxation, information that cannot be obtained directly in the frequency domain but that is of fundamental importance to a theoretical understanding of such relaxation. Recent picosecond studies of intramolecular rearrangement times have shown deviations from simple diffusionlike behavior that may be due to viscoelasticity. These techniques should continue to provide new information, particularly as they move from the developmental stage to the point where they can be more readily applied to a wide variety of systems over a range of physical conditions.

Experimental techniques that probe collective dynamics in liquids include dielectric relaxation, ultrasound and viscoelasticity measurements, light scattering, flow and acoustic birefringence, and neutron scattering. Much of the current work using these techniques is aimed at studying collective motions on time scales where macroscopic hydrodynamics no longer applies; here the details of the intermolecular forces and collisional dynamics become more important. Such studies are the result of technical advances that have enabled measurements to be taken at higher frequencies or shorter times, and the extension of measurements to lower temperatures and higher viscosities, where the characteristic relaxation times are slower, bringing faster processes into experimentally accessible regions. In particular, it has been found that, in contrast to the situation at higher temperatures and lower viscosities, many of the collective relaxation processes in viscous fluids are highly nonexponential, a phenomenon for which there is still no convincing theoretical interpretation. Since generalized hydrodynamics provides a theoretical framework in which data obtained using different techniques can be analyzed in a consistent fashion, it is especially important that data be obtained over a wide range of physical conditions using complementary techniques, for instance, light scattering and acoustic measurements. Technical developments will continue to provide new and better information. For example, new advances in time-domain dielectric relaxation have extended the applicability of this technique to shorter times or higher frequencies. Nonlinear optical techniques should be of benefit in studies of collective as well as single-particle properties. Newly developed optically induced transient grating experiments (laser-induced phonons) allow for the generation and study of very-high-frequency ultrasonic waves. The low-frequency analogue of Raman gain spectroscopy could provide an attractive alternative to Fabry-Perot interferometry for the study of dynamic light-scattering spectra of viscous fluids, since the inherent frequency resolution is much higher, enabling the study of slower processes and more viscous fluids.

There are many ways to model the physics of a liquid in order to obtain predictions of its dynamical behavior. One important approach used today is kinetic theory. Here one follows sequences of molecular collisions and determines spectra and transport coefficients as a direct consequence of the collisional history of the molecule. The unified collision-based theory of fluids began with Boltzmann's classic treatment of gases (1873) and was extended to dense gases by Enskog

(1922). Recently, Enskog's approach has been systematically generalized so that it can now be used to treat systems approaching liquid densities. In the liquid regime, Enskog's picture of uncorrelated molecular collisions is simply inadequate. Several workers have made significant revisions in the basic framework of the Enskog theory in order to accommodate the effects of correlated sequences of collisions. The cage effect in liquids, for example, arises when molecules, say 1 and 2, are forced by molecule 3 to collide. Thus, molecule 3 cages molecules 1 and 2. The consequences of such recollisions are profound, even going so far as to undermine the usual density expansion approaches used to calculate transport coefficients. Most of the emphasis in liquid-state kinetic theory has been on smooth, hard sphere systems. For nonspherical molecules (basically all molecules in nature except the inert gases, liquid metals, and a few other exceptions), the state of kinetic theory is much more primitive. Only recently has the Engskog theory of nonspherical particles been applied to condensed-matter dynamics, and there it yields unsatisfactory and inaccurate predictions of the transport coefficients owing, perhaps, to the omission of correlated recollisions. The understanding of the properties of rigid nonspherical molecules is just in its infancy.

Our discussion up to this point has centered on rigid molecules whose dynamics can be treated using kinetic theory. The study of the dynamics of small, flexible molecules, such as the alkanes, is also interesting. The intramolecular rearrangements that take place in such molecules are primitive models for chemical reactions, and there has been renewed interest in determining the rates at which flexible molecules change shape and how such changes in shape affect properties involving overall rotation and translation. Historically, there has always been an interest in small-alkane dynamics, but earlier approaches dictated the motions by fiat, and thus provided few fundamental insights into molecular conformational dynamics. Today, one derives the equations of motion from Newton's laws, and then follows the time evolution of the system in order to determine how energy is transported through the molecule, the temperature of individual bonds, and in general how the molecule moves as a result of collisions with the solvent.

Molecular dynamics (MD) computer simulations have, since the 1950s, continued to point out interesting phenomena in liquids and, sometimes, even hints at their explanation. Perhaps the most important developments in the past 5 years involve the applications of MD to (1) nonlinear phenomena, as seen through nonequilibrium molecular dynamics (NEMD) and (2) the dynamics of polyatomic molecules. In the

NEMD technique, one applies an external disturbance to a collection of, say, 500 molecules in a box. The disturbance might be a shear gradient. One then observes the induced momentum flux in the fluid as a consequence of the shear; the proportionality between the flux and shear gradient defines the shear viscosity. This technique provides a calculation of the shear viscosity and other transport coefficients that is more efficient than direct MD. Shear NEMD calculations have demonstrated that the shear viscosity has a square-root dependence on the magnitude of the shear gradient and on the frequency of shear (Figure 9.2), observations that raise important conceptual questions in the theory of fluids. Computer simulations of fluids composed of nonspherical molecules have played a similar role by providing details of molecular dynamics inaccessible to experiment. For example, it has been observed that a characteristic feature of rotational dynamics in condensed phases is an oscillation in the angular-velocity time-correlation function. Experimental transport coefficients, which are

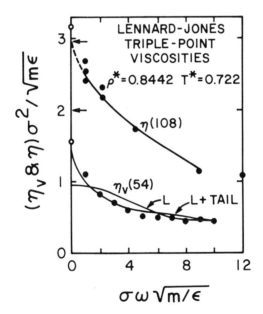

FIGURE 9.2 Shear (upper) and bulk (lower) viscosities for triple-point Lennard-Jones fluids as functions of a dimensionless strain rate. The experimental viscosities for liquid argon are indicated by horizontal arrows. The typical square-root behavior of these viscosities is responsible for the cusp in the Newtonian zero-strain-rate limit. [From W. G. Hoover, D. J. Evans, R. B. Hickman, A. J. C. Ladd, W. T. Ashurst, and B. Moran, Phys. Rev. A *22*, 1690 (1980).]

given by the time integral of the correlation function, do not readily see this feature. This oscillation indicates that the backscattering or caging mentioned in connection with molecular translation is also crucial to the understanding of the dynamics of molecular rotation in a liquid. In other words, correlated sequences of collisions must be understood in order to predict liquid properties.

Colloidal Systems—Soap Solutions

Solutions of soap in water are a familiar part of our everyday life; they also account for several multibillion-dollar industries involving detergent action, drug delivery, and oil recovery. Nevertheless, little is understood on a fundamental level about the many different ways in which soap molecules are aggregated in aqueous solvents. Sometimes they go into solution by means of the formation of spherical clusters of molecules; other times these aggregates—or micelles—are distinctly nonspherical, e.g., rodlike or disklike in shape. At high enough concentrations suspensions of these rods and disks are observed to transform themselves into stacks of infinite cylinders and lamellar sheets. In many instances, particularly on the addition of salt or alcohol or another soap species, intermediate phases appear in which the rods and disks remain small but tend to align along a single direction. In each of these various states, of course, the solution of aggregates displays markedly different mechanical, flow, and solubility properties.

To explain these features it is not sufficient to apply the usual theories of colloidal suspensions. This is because—unlike the cases of metal grains or biological macromolecules, say—the interacting particles in soap solutions are aggregates of molecules that do not maintain their integrity. Instead, any change in thermodynamic parameters such as temperature or concentration results in a reorganization of the clusters into a new distribution of sizes and shapes. Furthermore, since the particles themselves undergo change, so do the forces between them. Accordingly, a statistical-mechanical treatment of the bulk properties of concentrated soap solutions must necessarily confront explicitly the coupling between micellar growth and interactions. Similarly, the experimental study of these systems is also more problematic than that of ordinary colloidal suspensions.

During the past decade, much progress has been made in developing the theoretical concepts necessary for understanding micellization in aqueous soap solutions. Particular emphasis has been on accounting for the various preferred curvatures assumed by different aggregates, relating these different geometries to molecular shapes, degree of ionization, and overall soap concentration. Just as importantly there have been dramatic advances in the resolution of neutron, x-ray, and light-scattering experiments relevant to determining the microscopic structures of these systems. Furthermore, much effort has been devoted to the microemulsions that form on addition of oil to micellized solutions of soap in water. In most cases a cosurfactant (e.g., another soap molecule or an alcohol) is necessary to stabilize these oil/water dispersions. Thus one is dealing in general with at least a four-component, concentrated solution that shows a dramatically rich polymorphism at room temperature. These phases are also often characterized by extremely low interfacial tensions ($\leq 10^{-3}$ dyne/cm), making them of great interest for enhanced oil solubilization as well as for studies of fundamental thermodynamics and critical phenomena. It appears that the new and diverse phenomena displayed by these systems will continue to provide many fruitful challenges to our current ideas concerning the effects of dimensionality and symmetry on phase transitions and equilibrium structures.

LIQUID CRYSTALS

What Are Liquid Crystals?

The name liquid crystals covers a broad category of materials exhibiting molecular organization and macroscopic symmetry intermediate between the total disorder of an isotropic liquid and the order of perfect crystals:

Nematic phases are ones in which the centers of the molecules making up the material are more or less randomly arranged as in an ordinary liquid, while the orientation of the molecules exhibits long-range order. For instance, rodlike molecules are oriented with their long axes parallel to one another or disklike molecules with their plane surfaces parallel. These phases flow like ordinary liquids but exhibit the anisotropic optical, electrical, and magnetic properties usually associated with crystals.

Smectic phases are layered systems, so that they resemble crystals in having periodic order in one direction (the layers), while retaining some degree of disorder within the layers. There are many subtle variations on this theme, the simplest of which is the smectic A phase with liquidlike disorder in each layer.

Within these two general classes of partially ordered materials there are several subclasses, which is one source of richness in the field.

Another source of richness is the tremendous variety of materials that exhibit these kinds of ordering:

Small organic molecules containing roughly 40-100 atoms are commonly the rodlike units that make up nematic phases. Equally often these same materials exhibit smectic phases with the rods packed into layers, the rod axes being either perpendicular to the layers (smectic A) or at an oblique angle to the layers (smectic C).

Amphiphilic systems, based on molecules in which one end is oil soluble and the other end is water soluble, usually organize themselves into basically layered structures with the oil-soluble parts in one plane of the layer and the water-soluble parts in another plane. By simply stacking up the layers one can build smectic phases, but more complex structures can be achieved too. For instance, the amphiphilic layer can be rolled into a cylinder, and arrays of these cylinders suspended in water or oil can form a nematic phase.

Colloidal systems of objects larger than single molecules, for instance virus particles suspended in water, can form liquid crystals. Several viruses are rodlike in shape and make nematic phases.

Polymers consisting of rodlike molecular units strung together end to end, or attached like the teeth of a comb to a flexible molecular string, often exhibit nematic ordering. As one might expect, the mechanical properties of these systems are different from those of liquid crystals made of small molecules.

Biological subcellular structures such as cell membranes exhibit molecular organization and other properties similar to various liquid-crystal phases. In some cases these are really liquid crystals, while in other cases the structural complexity of the biological systems exceeds that of a liquid crystal, so using the terminology of liquid-crystal physics to describe the biological system is more an aid to thinking than a real physical description.

Why Are Liquid Crystals Interesting?

Although liquid crystals have been known since 1888, it is not unfair to say that the last decade has brought a surge of interest in their physical properties. Clearly one source of the fascination with liquid crystals is the variety of substances that exhibit these phases. That stimulates one to look for the unifying principles responsible for the similarity of behavior of widely differing systems. At the same time, the variety of different liquid-crystal phases exhibited by similar small molecules leads one to try to understand the subtle differences in molecular properties that lead to different kinds of ordering (smectic versus nematic, for instance). This has led to both fundamental theoretical research on molecular ordering and interactions and the development of new materials, in a kind of molecular engineering to achieve systems with specific properties.

A second source of fascination is the large number of unusual macroscopic phenomena found uniquely in liquid-crystal phases. These include dramatic changes in the macroscopic structure of a sample induced by a magnetic or electric field or by flow. For instance, in an initially undistorted single crystal of a nematic, an applied field may produce a periodic stripelike structure. This rather complex response to a simple applied force is striking. Without going into the detailed analysis of any of the myriad of cases in which something like this occurs, one can say that it results from the anisotropic nature of the coupling of the applied field to the liquid crystal. As soon as the liquid crystal begins to respond to the field, which usually involves a change in orientation of the molecules, the change of orientation results in a change of the strength of the coupling of the liquid crystal to the external field. This is an example of a nonlinear response that often leads to complex structural changes in a sample submitted to rather simple external forces.

These phenomena have proved challenging and stimulating in a number of ways. Understanding them and learning to produce and control them has led both to a deeper understanding of liquid crystals and nonlinear phenomena and to some interesting applications of these materials. Most of these macroscopic phenomena involve changes in the optical properties of the sample, much larger changes than are ever observed in ordinary crystals or liquids. As a result, most of the applications of liquid crystals to date are to various display devices, such as the digital readout of a wrist watch or a calculator; applications to television-type displays are in the near future.

Finally the changes of state exhibited by liquid-crystal-forming materials have been interesting. These include changes between an ordinary liquid or solution and a liquid crystal, as well as changes between various liquid-crystal phases. A number of these changes of state fall in the category of continuous phase transitions, which may exhibit critical phenomena owing to fluctuation effects. In addition, the rich variety of phase changes offered by these materials in films as thin as two monolayers has presented unique challenges in the fields of two-dimensional melting and the ordering of defects. These phenomena have been the subject of intensive research in recent years, and liquid crystals have provided a rich testing ground for theoretical ideas as well as a challenging array of phenomena to stimulate new ideas. In fact, one of the most common liquid-crystal phase changes, from the nematic to the smectic A phase, has still not been completely understood.

Major Advances

Liquid-crystal displays have become the dominant form of display in applications requiring low power, portability, or operation in a wide range of lighting conditions but are limited to cases in which only a small to moderate amount of information has to be displayed. Thus, they are widely used in wrist watches and calculators but not to replace cathode-ray tubes in computer terminals or television sets.

This achievement of research in liquid crystals has resulted from a combination of important contributions from various sources. First, the liquid-crystal displays now used are based on the twisted nematic polarization switch effect, an electric-field effect in which the internal orientational structure of the liquid crystal sample is changed in a way that rotates the polarization of light passing through it. Understanding the macroscopic phenomenology of this effect and making it reliable for practical application involved correct preparation of sample surfaces, development of ways to prevent the formation of defects during device operation, understanding the dynamics of the liquid crystal's response to electric fields, and understanding the rather complex optics of the device.

Second, new materials with the properties necessary to make these devices practical had to be developed. These materials had to combine properties such as a wide nematic temperature range around room temperature, chemical stability for a long life, as well as ideal optical, electrical, elastic, and viscous properties. The development of successful materials for this application has been a major achievement resulting from close collaboration between physicists and chemists.

In addition, a number of related technological developments were needed, including sample sealing methods, surface treatment techniques, and electrical signal-handling techniques.

This is a specific example of an aspect of liquid-crystal science that has been essential to the field—the interdisciplinary nature of the subject.

A second major achievement of the field has been the understanding and development of a number of new states of molecular organization and ordering. Again, this has required close cooperation between chemists and physicists and the interplay of theory and experiments.

The variety of partially ordered states of matter that can properly be called liquid crystals is remarkable. This subject has attracted some of the brightest researchers and is in a state of intense development now.

There are numerous other outstanding achievements in this field, some of which would require detailed technical discussion to be described meaningfully. These include the development of ultra-high-strength fibers spun from liquid-crystal materials and the discovery of ferroelectric liquid crystals that have a spontaneous electrical polarization.

OPPORTUNITIES FOR FUTURE WORK

One of the major areas of the physics of the liquid state where little real understanding exists is that of fluids away from, and especially far away from, equilibrium. Although some partial and fragmentary knowledge is available, neither the average properties of such fluids (such as the flow and density fields) nor the fluctuation phenomena about the average and their correlations are well understood. There are two aspects to this:

1. A fundamental microscopic theory of dense fluids not in equilibrium is not available. Since progress on this problem is slow, it is not a fashionable topic, which does not mean, of course, that it is not important. So far, only an approximate theory for a fluid of hard spheres has been developed, and some modest attempts are under way to generalize this theory to more realistic fluids. However, we are still far from any detailed microscopic understanding of the nonequilibrium properties of real fluids. New approaches, both theoretical and experimental, for dealing with this problem are being developed, and this development should be encouraged. Among the new possibilities for the experimental study of liquids out of equilibrium one should mention laser spectroscopy, improved neutron spectroscopy with the new spallation sources (such as the Los Alamos Neutron Scattering Center), and synchrotron radiation. The use of synchrotron radiation could also help in clarifying the behavior of chemically reacting mixtures, about which more basic knowledge, both theoretical and experimental, would be highly desirable.

2. A fundamental macroscopic understanding of the behavior of fluids far from equilibrium based on the nonlinear equations of hydrodynamics, such as the approach to chaos or turbulence, will certainly remain an important and fashionable topic for research in the future. There is an interesting connection here with the behavior of some

chemically reacting fluid mixtures, in which diffusion also occurs (as, for example, in the Belousov-Zabotinskii reaction), and clarification of the relevant basic equations, aided by recent advances in our understanding of critical phenomena, will be of great theoretical and practical importance.

An understanding of non-Newtonian fluids (e.g., their rheological properties) from a more physical rather than from an abstract mathematical point of view is being gained. However, a real, basic understanding of such fluids, sometimes necessary for shrewd practical applications, is still lacking. The properties of liquid crystals, glasses, polymers, and gels, for example, are being studied from various points of view. All this work should be encouraged and supported. However, there is a great need for a unification of the various approaches. In polymer science, for example, the theoretical-physics approach and the chemical-engineering approach are not at all compatible, and this slows progress in the field.

For both transport and equilibrium properties, and for the inclusion of three- and higher-particle effects into these calculations, a more concentrated effort is needed in the future. A particularly interesting opportunity is the further development of models for polyatomic fluids, and the development of theories of nonuniform fluids, in the context of which liquid against solid interfaces provide an important example. In view of the strong connection between the liquid-interface problem and large-scale commercial chemical-engineering processes (including, for example, catalytic processes) it seems clear that more emphasis and greater support should be devoted to experimental measurements as well as to the microscopic understanding of liquids and their mixtures, both uniform and nonuniform.

Liquid-crystal research is at an interesting point in its evolution. There has been sustained activity in the field on a number of fronts for the last 15 years. In spite of the advances made, it is still clear that the number of new questions being encountered outweighs the number of problems solved.

In the area of liquid-crystal displays, which has served as a fundamental motivating force for much research, there is the potential for major new developments. The currently successful twisted nematic displays are capable of practical application only to situations requiring display of a relatively small amount of data. This is because the display must constantly be refreshed: it has no internal memory. Much effort is being devoted to the development of a display with intrinsic memory in addition to all the desirable features of the twisted nematic displays.

Another issue is speed: the current displays can be updated only relatively slowly. The combination of greater speed and intrinsic memory would make television applications of liquid crystals especially feasible. Some of the most promising research in this area now concerns the use of ferroelectric liquid crystals, one of the striking discoveries of the last decade that has not yet been fully developed. As with the twisted nematics, success in this area will depend on the cooperation of physicists and chemists and on the development of knowledge in a number of areas that are currently not well understood, such as the interaction of smectic liquid crystals with surfaces.

There is currently rapid development in the understanding of new kinds of molecular ordering and phase transitions. Much of this has been associated with high-resolution x-ray studies of the various smectic phases, in combination with other studies such as optical and NMR experiments. The availability of national synchrotron-radiation facilities has played an important role in this development.

Whereas in the recent past most of the emphasis has been placed on the study of liquid crystals based on small organic molecules, now considerably more effort by physicists is being devoted to liquid crystals formed by amphiphilic systems, polymers, colloidal suspensions, and biological substructures. This broadening of interests is leading to a number of new discoveries.

As in any rapidly developing field, of course, there are many interesting questions encountered that go unanswered as a particular area of the subject is explored. In this sense, even within the generally well-studied aspects of liquid crystals, there are still many opportunities for productive research.

10

Polymers

INTRODUCTION

Polymers are macromolecules made up of long sequences (thousands) of small chemical units repetitively attached by strong chemical bonds to form chains or other structures. Differences from one type of polymer to another are due not only to the chemical nature of their constituents but also to their physical arrangement. Small structural differences, such as branching or cross-linking, can produce profound differences in properties.

Research in the field of polymer science is a massive endeavor in the United States and abroad. Polymers possess a range of properties, often unique, that have proved to be adaptable to a wide variety of uses. The production of polymers in this country as measured by volume exceeds that of steel.

The investigation of polymers is diverse, requiring interdisciplinary efforts of physicists, chemists, materials scientists, biochemists and biophysicists, and chemical and mechanical engineers. Research in the field is currently vigorous. In the past 10 or 15 years new instrumental developments, e.g., small-angle neutron scattering; Fourier transform infrared spectroscopy; solid-state nuclear magnetic resonance (NMR); light, x-ray, and electron scattering; electron microscopy; new surface probes; computerized instrumentation; and computer simulation, have had a large impact. Paralleling this have been theoretical breakthroughs in problem areas central to the field, e.g., polymer disentanglement, the excluded volume problem, gelation, and nonlinear mechanics. Another factor enlivening the field is the uncovering of new materials and properties. Examples are semiconducting and conducting polymers, piezoelectrics, liquid crystals, block copolymers, high-strength extruded materials, immobilized enzymes, and polymeric membranes.

RESEARCH PROBLEMS

Amorphous State—Solutions and Melts

The dilute-solution state is one in which attention can be focused on the behavior of individual macromolecules. For the most part polymer chains form loose coils, and it has been useful to draw an analogy between the path of such a chain and the path a walker might follow wandering randomly through space. However, there is an important difference. The walker can freely recross his path, but the polymer chain cannot cross parts of its path already occupied. This is referred to as the excluded volume, or self-avoiding walk, problem. A coiled molecule expands, on average, to decrease regions of overlap. Whereas the average end-to-end distance of a random walk of N steps goes like $N^{1/2}$, that of a self-avoiding walk goes like N^v with $v \approx 0.6$. The appearance of a characteristic exponent is reminiscent of critical phenomena, reviewed elsewhere in this report (see Chapter 3). In that chapter there is a discussion of magnetic models and their differences according to the dimensionality of the spin. Formally it is found that the polymer problem is in the same universality class as a magnet of zero-spin dimensionality This may make no physical sense for magnets, but it illustrates the importance in modern theory of limits defined only mathematically.

Once the theory of critical phenomena was applied to polymer problems it proved capable of describing a vast array of observations, both static and dynamic. This includes the description of a state unique to polymers, semidilute solutions. Dilute solutions are those in which the units are widely separated. For polystyrene of molecular weight 10^6 g/mol (about 10,000 styrene monomers per molecule) a solution of about 0.1 percent polymer has monomer units widely separated, but the polymer molecules as a whole are beginning to overlap each other, i.e., are not dilute. In this semidilute condition, a screening of the excluded volume interactions between monomers on the same chain develops. Screening is another critical phenomena concept (cf. Chapter 3). From an experimental point of view, common polymers may not be long enough for the semidilute theory to apply fully over a large range of concentrations; i.e., they may always be in what is termed a crossover region between dilute and semidilute. It is hoped that progress in handling such crossover effects will soon be made. Polymers provide many other manifestations of crossovers that may help to delineate the effects.

An extremely valuable advance in investigating polymers occurred some 10 years ago with the development of small-angle neutron scattering. It opened up the possibility for studying the properties of single-polymer molecules in condensed states, when they are permeated by other molecules of the same kind. This is done by labeling some of the molecules through replacement of hydrogen atoms with deuterium. It was quickly confirmed that, in the melt, random-walk statistics apply to single-polymer chains. Since then many other results, some quite surprising, have been obtained (see below).

The flow properties (rheology) of polymers are rather unusual, exhibiting long-term memory, viscoelastic effects, and nonlinearities. These are due to the fact that polymer systems, melt or semidilute, are entangled masses. When a strain is induced in a polymer melt the individual molecules are distorted and continue to exert a force (stress) resisting that strain, until the molecules have moved out of the strained, entangled mass and have relaxed to an equilibrium entangled condition. The quantitative description of the dynamic entanglement problem has recently been achieved in a marvelously simple way. Consider a single, long-chain molecule. The molecules surrounding it can be considered, effectively, to form a tube. Think of a snake in a tube as long as itself. If the tube is distorted these distortions are transferred to the snake. That distortion is experienced on some part of the snake as it moves out of the tube until it has fully escaped. Of course, escape is only possible by moving along the tube, lateral motions being prevented. Because of this analogy the proposal is called the reptation theory. It has been successfully applied to describe diffusion, rheology, relaxation of rubbers, healing of cracks, crystallization from the melt, and phase-separation dynamics. There is much experimental activity aimed at testing the applicability and limits of the theory and at developing new refinements, extensions, and uses.

To this point we have discussed problems involving properties of systems where the most important aspect is the chain character of the molecule, and the detailed structure and motions on the atomic scale (nanometers) are of peripheral importance. There are many characteristics of structure, packing, and dynamics on the smaller scale that affect polymer properties. Modern tools have advanced the science enormously in a decade. These include NMR, fluorescence spectroscopy, Raman and infrared spectroscopy (especially Fourier transform infrared spectroscopy, or FTIR), wide-angle neutron scattering, neutron spin-echo spectroscopy, and computer simulation. Traditional methods have been important, too. Light has been shed on how the small units of the molecule pack and manage to move, given that they are restricted by being only pieces of a larger molecule. Structure and properties have been correlated as a step toward a fuller understanding of their relationship. An aim, actually achieved to some extent, is to tailor-make materials with desired physical properties by controlling the molecular or physical structure.

Glass

The glassy state is extremely common in polymeric materials. The usual glassy brittleness has been circumvented by blending and grafting glassy polymers with rubbery particles in complex ways that are not fully understood. Let us concentrate here on the pure glasses. In general one can say that the theoretical foundations for describing the glass are primitive compared with other physical states of matter. There is substantial, but indirect, evidence that the explanation for some properties involves atom-sized bits of free volume in the glass and that motion is only possible in association with this free volume.

What is a glass? It is a disordered material in which the times are long for relaxation back to equilibrium following a change of physical conditions (e.g., temperature or stress). These times may be seconds or centuries. It was once hoped that a description could be achieved in terms of only one extra fundamental parameter that had not relaxed. Recent evidence is that this is not so. However, there seems to be a universal function that the slow relaxations obey. If the system is driven (or normally fluctuates) out of equilibrium, it returns according to the formula $\exp[-(t/\tau)^\beta]$, where t is the time and τ and β are parameters. Unfortunately this is not a mathematical expression that is frequently encountered in physics, so little idea exists of what the underlying mechanisms are.

Elastomers, Gels, Cross-linked Networks

One of the outstanding, early achievements of polymer physics was the development of a working picture for rubber elasticity. A rubber (gel, elastomer) is formed from a nonglassy, amorphous polymer when the molecules are tied together by a few cross-links per molecule.

Basically, macroscopic deformation of the specimen is reflected in a distortion in the positions of the cross-link points. This, in turn, decreases the number of ways the chains between the cross-links can arrange themselves (technically speaking, decreases the entropy). Decreasing entropy takes work, just as increasing energy does (as in stretching a spring), hence the resisting force of a stretched rubber. It was long ago realized that idealized calculations of the elastic force were imperfect, but difficulties in preparing well-characterized samples inhibited (and still do) the test of refined models, for instance those that attempt to account for entanglement. Neutron scattering provides a powerful tool for probing this problem, but the early results do not agree with any of the theories. Explanations may involve a deeper analysis of the topology of the network and its reaction to strain (e.g., the unfolding of three-dimensional pleats).

In the course of the process whereby a collection of single molecules is transformed into a totally connected network by progressive cross-linking, there is one critical amount of cross-linking that leads to the first appearance of a cluster spanning the sample. This is the gel point. The properties of systems near this condition are well described as critical phenomena. Ramifications of this description are being pursued.

Polymer Crystals

Looked at on the scale of atomic spacings (nanometers), polymer crystals exhibit the regular characteristic of small-molecule crystals. Looked at on larger scales, the differences are legion. Each crystal that forms tends to be surrounded by amorphous material. Fractional crystallinity might typically be in the 20 to 70 percent range. The crystals are commonly lamellar (plateletlike) in shape, with a thickness of 10 to 20 nm and much larger in other directions. The molecules stretch back and forth between the lamellar faces. The polymer chains form high-energy folds at each face and usually re-enter the crystal either adjacently or in nearby positions. The crystal would be more stable if it were thicker (fewer folds per molecule), so this is not the equilibrium state. The crystals form this way for kinetic reasons. To form a more stable crystal would take so long that it does not occur ordinarily. Thus one is faced with the challenge of deciphering the details of the nucleation bottlenecks to growth in order to understand and predict properties of the crystal. The number of proposed growth processes applicable under various conditions has recently increased. After a period of quiescence this problem has received considerable attention of late because of unexpected results that emerged when the state of individual molecules was examined with neutron scattering.

Viewed on a still larger scale, the lamellae frequently arrange themselves in a spherulitic pattern of bifurcating, radial fibrils, often twisting as they grow. Note that the presence of a twist implies a handedness, which is not inherent to the symmetry of the molecule. Recently, attention has focused on how that broken symmetry is introduced and propagated.

There are other varieties of crystal morphology, such as fibers formed on drawing or the forms observed when growth is in contact with certain surfaces. Also it appears possible to grow extended chain crystals under pressure, which may be related to liquid-crystallike ordering.

Electrical Properties

Traditionally, attention has focused on many polymers because of their properties as electrical insulators. Recently, however, the attention of numerous physicists, many new to the field, has turned to macromolecules because of the discovery of interesting electrical properties in certain polymers. An example is polyacetylene, which when pure is a semiconductor but can be doped into the range of metallic conduction. The polymer consists of a chain of carbon atoms with hydrogen atoms attached to each. In a small ring of this nature all the carbon-carbon bonds would be equivalent, and a half-filled metallic band would be formed. In a large ring or a long chain, however, the polymer lowers its energy by displacing atoms to create alternating single and double bonds between carbon atoms, which gives rise to insulating electronic bands. There are two degenerate structures formed by this dimerization, each formed from the other by the interchange of the double bonds. Thus either bonds 1, 3, 5, . . . or 2, 4, 6, . . . can be double. Occasionally one gets transitions between the odd and the even patterns. The resulting walls that separate domains of the two degenerate structures are mobile, and are associated with excited electronic states that spread over some 15 atoms. They are called topological solitons. The dopant modifies the electronic state of the soliton, creating charge donors and acceptors. One of the most remarkable properties of these solitons is that the relation between their spin and charge is reversed from the usual situation; i.e., a soliton with charge 0 has spin 1/2, while a soliton with charge $\pm e$ has spin 0. These appear to be the actual stable states for mobile neutral defects and charges in polyacetylene. Extensions of this idea to other cases has shown that the charge partitioning can be even more pathological, yielding net fractional charges on solitons that can be rational or irrational. These electronic states can be investigated by various spectroscopic techniques. Recently efforts have been made to look at the properties of polyacetylene in a way that integrates physical, structural, and chemical (bond arrangements and transitions) aspects.

Another material with interesting electrical properties is polyvinylidene fluoride, which is ferroelectric, piezoelectric, and pyroelectric. Microscopically these important materials properties arise from dipoles (on the monomer unit) that can be oriented by subjecting the polymer films to high electric fields at elevated temperatures (well above the glass transition). The possibility of building regular dipoles into polymer structures and creating a wider class of such materials is certainly an exciting direction for the future. The combination of the mechanical properties of polymers combined with piezoelectric and other physical properties offers the promise of a variety of technological applications.

Other Polymer Properties

There are other types of polymers, and properties of polymers, no less interesting and important than those just described, that space does not permit us to discuss in any detail.

1. Some polymers form liquid crystalline phases as an outgrowth of the rigidity of the backbone or substituent groups attached to the chain. This ordering leads to some high-strength materials.

2. Commonly polymers are blended to form useful materials that are either true mixtures or intimately associated microphases.

3. Block copolymers are made up of two or more chains attached in the same molecule. Phase separation may occur, but only microdomains can form because of the chemical connection between the separated units.

4. Some polymers contain ionizable groups. Their structure, in solution or bulk, is strongly influenced by Coulomb forces.

5. The surface is the face a polymeric phase presents to the world. This surface may be the natural one or one modified deliberately or through aging. Tremendous progress in surface science has been utilized by polymer researchers.

6. Last, but by no means least, mention should be made of some of the problems associated with bipolymers: organization, kinetics, function, compatibility, mechanical properties, and transitions.

The excitement of the polymer field is an outgrowth of the diversity of properties that these materials exhibit, a list of properties that keeps growing by discovery or design.

OPPORTUNITIES

The following list highlights several important areas of polymer physics in which significant progress may be expected (or at least hoped for) in the next few years. It is intended to be representative and not comprehensive:

1. Experimental and theoretical efforts that contribute to a fundamental understanding and/or phenomenological description of glasses, including polymeric ones. The nature of relaxational motions and how these relate to ultimate strength.

2. Understanding of crazes formed during failure and application of that knowledge to the toughening of glasses. (Crazes are microcracks caused by environment and/or mechanical working.)

3. A broader development of the reptation idea to the description of processes influenced by entanglements. A better connection between a fundamental description of entanglements and the effective tube description. Attack on a few persistent disagreements between theory and experiments.

4. Characterization of polymer properties under conditions corresponding to crossovers between asymptotic regimes, such as solution concentrations between dilute and semidilute.

5. Development and utilization of molecular tags that can be attached to macromolecules. These tags should have properties, such as spectra, fluorescence, or scattering cross section, that make them easier to observe than the polymers themselves. They should reflect the polymer's phase structure or dynamics. The question of the degree to which these properties are modified by the tags should be clarified.

6. Transport of low-molecular-weight molecules through polymers. Use of low-molecular-weight molecules as probes of polymer properties.

7. Various studies of polymer dynamics employing high fluxes of synchrotron radiation or pulsed neutrons.

8. Definitive characterization of the polymer-fold surface of crystalline lamellae.

9. Description of the various polymer crystal nucleation, growth, and aging processes, to explain such observations as curved crystals, spherulitic growth of twisted fibrils, and thickening of lamellae during annealing.

10. Rheology of liquid crystalline polymers. Ordering effects of flows and electrical fields on these materials.

11. Mechanisms of charge conduction along and between conduct-

ing macromolecules of various types. The role of dopants. Control of the nonelectrical properties (e.g., solubility, morphology, strength, degradation) of these materials.

12. Rheology of polymer blends (those that are fine dispersions), composites (polymer matrices with particulate or fibrous inclusions), thin films, and block copolymers.

13. Modification of polymer surfaces so that their chemical and physical properties (e.g., adhesion, biocompatibility, catalysis, reactivity) differ from those of the bulk.

14. Kinetics of phase separation.

15. Understanding of the forces and factors governing polymer miscibility.

16. The role of entanglements in rubber elasticity.

17. Development of an understanding of the physical factors governing the three-dimensional ordering of bipolymers, sufficient to make quantitative predictions of that ordering *in vivo*, and disruption of order by solvents, heat, and other agents. Description of how the ordering influences biological function, especially with respect to complex processes such as enzymatic action and membrane transport. Structure and function of membranes and their protein inclusions.

11

Nonlinear Dynamics, Instabilities, and Chaos

INTRODUCTION

A new area of condensed-matter physics has been defined as a coherent subfield in the past 10 years. When solid or fluid systems are forced away from equilibrium, they often become strongly nonlinear and as a result exhibit instabilities leading to chaotic or nonperiodic time evolution. In other cases, nonequilibrium systems develop patterned but nonperiodic spatial structures. From a general point of view, researchers in this field are concerned with the problem of predictability in disordered physical systems. Under what conditions can we hope to predict the future behavior of a system that is evolving chaotically, or the spatial structure of one that is highly fractured? Is any predictability preserved when a system behaves in an apparently irregular way? The process of obtaining answers to these questions should have a significant impact on fields of science extending far beyond condensed-matter physics.

One focus of experimental activity in this field has been the problem of understanding fluid instabilities (pattern changes) and turbulence, subjects of both fundamental interest and practical importance because of the simplicity of fluids and their ubiquity in nature. However, chaotic dynamics are increasingly being found in other areas of condensed-matter physics, such as superconducting devices and electronic conduction in semiconductors. Phenomena exhibiting complex

215

spatial patterns include the aggregation of smoke particles and the dendritic growth of crystals.

One of the defining characteristics of this subfield is a strong interface with the area of mathematics known as dynamical systems theory, which provides a language and set of concepts for understanding chaotic dynamics as a consequence of nonlinear models with only a few degrees of freedom. This approach to understanding irregular behavior is different from the conventional viewpoint of statistical mechanics. It represents an expansion of the theoretical tools available to condensed-matter physics. The eventual impact of these new mathematical tools will probably be greater than that of the specific physical systems to which they are currently being applied. Although the discussion in this chapter emphasizes nonlinear dynamics in the context of instabilities and chaotic motion in fluids and other condensed-matter systems, a much broader approach could be taken, in which the methods of nonlinear dynamics are applied to a wide range of interdisciplinary problems in plasma physics, astrophysics, and even accelerator technology.

Physicists working on nonlinear problems sometimes publish in engineering or mathematical journals, and new journals devoted specifically to nonlinear phenomena have been started. The interdisciplinary nature of nonlinear dynamics creates both opportunities for collaborations and special problems in structuring and funding research. These problems are different from those of the more established subfields of condensed-matter physics.

MAJOR ADVANCES

A New Paradigm

We now know that chaotic motion in physical systems that dissipate energy can arise from nonlinear effects alone: that is, apparently random behavior arises from the internal dynamics of the system rather than from irregular external influences. The first extensive discussion of chaotic motion of this type in a mathematical model was due to Lorenz in 1963. Since that time, a new mathematical language and way of thinking about nonlinear dynamics has been developed by mathematicians and (more recently) physicists.

The central concept is that of a state or phase space in which the state of a system is represented by a point, and its evolution appears as an extended trajectory. The trajectories form geometrical shapes or attractors in state space (see the discussion below in the section on

Dynamical Systems Analysis of Experiments) whose forms provide a way of characterizing the actual behavior of the physical system. For example, closed loops in state space describe periodic oscillations; attractors having the topology of a torus describe more complex oscillations termed quasi-periodic; and forms known as strange attractors describe chaotic motion. A strange attractor is a set on which trajectories wander erratically. They are generally extremely complex; a strange attractor often has the form of an infinitely folded sheet of infinite extent.

The description and explanation of the dynamics of chaotic systems is facilitated by a geometrical analysis of the shapes formed by the trajectories in state space. This dynamical systems analysis has been shown to be useful in unifying many diverse physical phenomena, because only a limited number of fundamentally different forms or types of attractors seem to be important experimentally. The development of tools and language for quantifying the properties of attractors has given physicists a powerful new way of thinking about the dynamics of nonlinear systems.

New Experimental Methods

New experimental methods were necessary to allow the ideas of nonlinear dynamics to be tested. Most importantly, laboratory computer techniques have been essential for experimental studies of complex dynamics. Automated data acquisition was required to obtain the large quantity of data needed for analysis. Computer methods have been needed not only to acquire data but also to analyze it. For example, space and time Fourier transform methods, which reveal the frequency content of the fluctuations, have been used extensively. Numerical methods of measuring the shapes and properties of strange attractors have been devised. Computer-enhanced shadowgraph images reveal flow patterns too feeble to be observed by ordinary photographic or visual techniques (an example is shown in Figure 11.1). These methods are still being refined, and are gradually narrowing the gap between the abstract mathematical ideas of nonlinear dynamics and the behavior of physical systems.

Routes to Chaos

How is a chaotic fluid flow reached as it is forced more and more strongly? The stress on the system is characterized by a dimensionless control parameter (Reynolds number or Rayleigh number, for example,

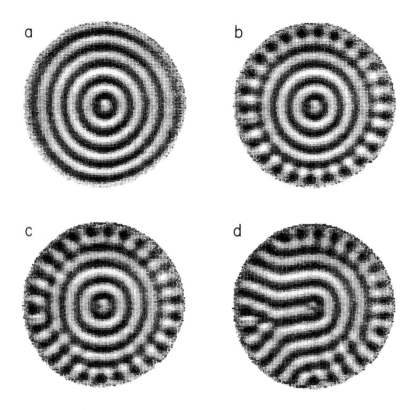

FIGURE 11.1   Computer-enhanced shadowgraph images of convective flow in a container of circular cross section. The concentric flow state (a) was rendered unstable by reducing the Rayleigh number from 1.2 times the critical value R_c to about 1.1 times R_c. Part of the resulting evolution of the flow field is given by b-d. (Courtesy of G. Ahlers.)

depending on the type of flow). Until about 1970, an endless sequence of instabilities was expected, each of which changes the spatial structure (pattern) of the flow or adds a new frequency of oscillation. In this picture, turbulent flow was essentially a complicated superposition of motion at many frequencies simultaneously. No qualitative change separated laminar and turbulent motion in this view. In 1971 a suggestion was made that turbulent flows could actually be modeled by strange attractors having only a few degrees of freedom. The chaotic dynamics of such models are dramatically different from a quasi-periodic flow with several frequencies and also quite different from chaos produced by many interacting variables. Considerable effort has been devoted to the testing of this hypothesis.

Over the next few years (1974-1982) many groups explored and characterized the various sequences of instabilities leading to chaotic fluid motion that are obtained as the control parameter is varied. A limited number of well-defined routes to chaos were identified:

1. Period doubling, in which the basic period of oscillation doubles repeatedly at a sequence of thresholds forming a geometric series. This phenomenon is illustrated in Figure 11.2.

2. Quasi-periodicity, in which chaos is produced by several interacting oscillations at incommensurate frequencies.

3. Intermittency, in which laminar and chaotic phases alternate in time.

These basic routes to chaotic motion were found both theoretically in simple models and experimentally in a limited class of fluid systems where the spatial pattern of the flow is fixed and the transition occurs continuously. We use the term chaos (or sometimes weak turbulence) to describe the resulting noisy time-dependent motion. The flow is not necessarily turbulent in the ordinary engineering sense because it may not contain eddies of many different spatial scales. Many of the same routes have also been observed in *p-n* junction oscillators, semiconductor transport phenomena, superconducting Josephson junctions, chemical reactions, and elsewhere. These routes to chaos have universal properties that transcend the characteristics of particular systems. The list given above is by no means exclusive; other routes occur in model systems and may be identified experimentally.

Dynamical Systems Theory of the Routes to Turbulence

The basic sequences of instabilities or transitions found experimentally are now fairly well understood theoretically. A set of theoretical methods known as renormalization-group techniques (developed originally to describe continuous phase transitions in equilibrium systems) were crucial in understanding the universal properties of the basic routes to turbulence. The applicability of these methods in the new context depends on the fact that the internal structure of strange attractors often becomes identical (more precisely, self-similar) on all length scales as the transition to chaos is approached.

Particularly noteworthy theoretical developments include prediction of the properties of the period doubling, quasi-periodicity, and

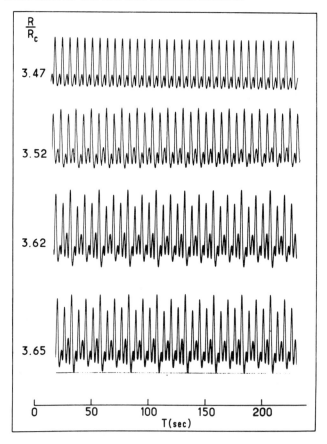

FIGURE 11.2   Sequence of period doublings leading to chaos in Rayleigh-Bénard convection, as a function of the Rayleigh number R (proportional to the temperature difference across the fluid layer). The temperature variation is periodic, and the period doubles repeatedly as R is increased. For example, at R/R_c = 3.62, a complete cycle includes four large peaks. (Courtesy of A. Libchaber.)

intermittency routes near the onset of chaos and the achievement of an understanding of the effect of external noise on these phenomena. These efforts have revealed new constants of nature and unexpected mathematics. The discovery that the routes to chaos have universal properties is a particularly exciting development and accounts in part for the speed with which nonlinear dynamics (applied especially but not exclusively to fluid motion) has been accepted as a part of condensed-matter physics.

Dynamical Systems Analysis of Experiments

The various routes to turbulence do not directly expose the properties of the strange attractors that characterize chaotic systems. In fact, the goal of understanding strange behavior has not yet been achieved, in part because it has only recently become possible to observe the attractors directly. Whereas earlier work emphasized the use of frequency spectra, one can now use geometrical methods to study the forms of strange attractors in phase space and to measure their properties. An example of the shape of trajectories in phase space near the onset of chaotic fluid flow is shown in Figure 11.3. When the

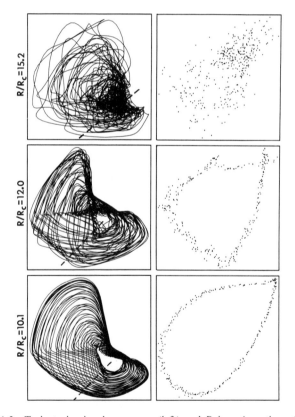

FIGURE 11.3   Trajectories in phase space (left) and Poincaré sections (right) for a rotating fluid below and above the onset of chaos. The flow is quasi-periodic in the lower panels and chaotic in the middle and upper panels. (Courtesy of H. L. Swinney.)

trajectories are shown in the two-dimensional phase space, they are smooth in the laminar state (lower left) and convoluted in the chaotic state (upper left). By observing the orbits in three dimensions and then allowing them to intersect a plane, the Poincaré maps on the right are obtained. The lower one (a closed loop) is the intersection of the torus with a plane. The upper one, where the flow is chaotic, is irregular because the torus has been replaced by a strange attractor. The middle case is just beyond the onset of the chaotic regime.

One of the most important properties of a strange attractor is its dimensionality, which can be fractional. This surprising property arises because it often consists of an infinite number of closely spaced surfaces, yielding a dimension greater than that of a single surface but lower than that of a solid object, i.e., a dimension between two and three. A second important property is the rate at which nearby trajectories diverge from each other along certain directions within the attractor. This property of exponential divergence of nearby trajectories or sensitive dependence on initial conditions is responsible for the chaotic behavior.

Preliminary measurements of these two properties in experiments have provided solid evidence for the basic correctness of the dynamical systems approach to understanding the onset of chaos in a limited class of fluid-flow transitions. We now know that chaotic motion in a system having 10^{23} degrees of freedom (i.e., a fluid) can be effectively described in some cases by models having only a very few degrees of freedom (variables).

Nonlinear Stability Theory

The dynamical systems approach described in the last few sections is particularly useful for systems in which the spatial patterns of the flow are fixed (for example, convection in a layer whose lateral dimensions are small enough). However, in many situations (including those of greatest practical importance) this is not the case: time dependence is accompanied by large variations in the spatial pattern of the flow. One widely applicable theoretical approach that has proven successful in describing pattern changes is known as stability theory. In this method, one examines the linear stability of the static solutions of the nonlinear equations to infinitesimal perturbations. A considerable amount has been learned about various types of secondary instabilities that modify the primary flow. Encouraging progress has also been made in understanding the effects of the system boundaries on the stability of the motion.

Pattern Evolution

Studies of the evolution of spatial flow patterns resulting from instabilities are an important step in understanding the onset of chaotic flow. During the past few years, an increasing effort has been devoted to studies of the evolution of flows, especially convective flows, by laser Doppler mapping and other methods. Particular attention has been directed to the effect of defects or dislocations on the evolution of the flow patterns.

Theoretical efforts have proceeded in parallel with the experiments. Most of the important qualitative phenomena seem to be predictable on the basis of two-dimensional model equations that are far simpler than the full three-dimensional hydrodynamic equations. These models are used to describe approximately the slow spatial variations of the basic flow structure. Numerical computations based on such model equations have been carried out successfully, and an example is shown in Figure 11.4. One remarkable development is that near the onset of convection, the evolution may possibly be described by the minimization of a quantity that depends on the concentration of defects, the amount of curvature in the pattern, and the orientation of the flow field with respect to the boundaries. Though evolution toward a minimum is surely not a general property of nonequilibrium systems, the usefulness of equilibrium concepts (approach to a minimum) in this context, even under restricted circumstances, is an important result.

Many other systems also show fascinating patterns and pattern selection. The growth of crystals from the melt exhibits complex dendritic or snowflake forms. The growth of soot particles shows somewhat less regular, but fundamentally similar, patterned growth. These spatial structures are, like the attractors of the fluid, strange objects having fractional dimension.

Various types of random motions can similarly be described in terms of objects with fractional dimension and can probably be understood using the language of dynamical systems. For example, the invasive percolation of a two-phase fluid in a porous medium, various aggregation processes, and the behavior of electrons hopping in a random material all show dynamically produced structures of fractional dimensionality.

Instabilities in Other Dissipative Systems

Instabilities and chaos occur in many other condensed-matter systems. For example, Josephson junction (superconducting) oscillators

FIGURE 11.4   Numerical simulation of convective-pattern evolution based on a 2-D model, starting from random initial conditions. The simplification of the flow into a pattern of textured rolls as time evolves is at least qualitatively similar to the phenomena observed experimentally.

show chaotic dynamics, a fact that may be important in device applications. Cooled extrinsic germanium photoconductors exhibit a series of progressively more complex instabilities with an increasing applied dc electric field. Niobium triselenide shows periodic and chaotic fluctuations in its conductivity that are related to the interaction of a charge-density wave with the crystal lattice. The spatial arrangement of adsorbed atoms on a periodic substrate and the electronic energy bands of a periodic crystal in a magnetic field both show complex structures that can be understood using the methods of nonlinear dynamics. The dynamics of chemically reacting systems are chaotic in some cases and in fact show a close correspondence with low-dimensional dynamic models.

Instabilities are also important in many aspects of the study of phase transitions between various states of matter. Progress has been made in studies of the nucleation of first-order phase transitions (i.e., the thermally activated initiation of a phase transition from a metastable state) and spinodal decomposition (phase transitions occurring from an initially unstable state). These processes are highly nonlinear and have been elucidated recently by light-scattering experiments on systems undergoing phase separation.

Nonlinear Dynamics of Conservative Systems

The nonequilibrium behavior of condensed-matter systems usually involves the dissipation of energy, and this means that the models of interest have attractors onto which the orbits in phase space converge. However, the nonlinear dynamics of conservative (nondissipative) systems has closely related properties and also gives rise to chaotic behavior. In this section, we briefly note that the history of this field is much older than is generally appreciated.

For 200 years after Newton, mathematicians and physicists sought to integrate Newton's equations and generally assumed that all Newtonian systems would eventually be found to be integrable or separable into equations for each degree of freedom and therefore to behave smoothly and predictably. However, Maxwell pointed out a century ago that the motion of a gas of hard spheres could quickly (in 10^{-10} second) be observably altered by a small perturbation and therefore could not be predictable. Early in its history, the probabilistic nature of statistical mechanics was realized not to be obviously consistent with an integrable and predictable Newtonian dynamics. Slightly later, in his study of the gravitational N-body problem, Poincaré proved that the energy is the only well-behaved constant of the motion for an isolated system, thereby casting grave doubts on the notion that Newton's equations could be integrated. Moreover, the wildly erratic character of the orbits of a system of three bodies led him to remark, "Determinism is a fantasy due to Laplace."

The nature of chaotic motion in conservative systems has been elucidated over the years by many distinguished scientists, including Birkhoff, Gibbs, and Hopf. An especially important concept is that of mixing, in which the flow of trajectories on a surface of constant energy in phase space mixes in the same way as the fluid particles in a stirred martini do. We now know that there are many conservative systems that are not only mixing but also truly chaotic in the sense that the final state is exponentially sensitive to changes in the initial conditions. Newtonian dynamics is thus quite rich; it contains the predictable systems of textbook physics as well as the fully chaotic, random systems envisioned by statistical mechanics. How can one go from one to the other? The much celebrated Kolmogorov-Arnol'd-Moser (KAM) theorem (1954) showed that small perturbations to integrable systems leave most orbits nearly unchanged (with suitable qualifications) while also creating a small set of chaotic orbits. It is this small chaotic set that can grow, as the conditions of the KAM theorem are violated, to yield fully chaotic motion.

The final and recent development in conservative dynamics comes via information theory. The basic notion here is that a chaotic system or orbit exhibits much greater variety, and therefore contains much more information, than does an orderly integrable system or orbit. This variety is sometimes quantified using the concept of metric entropy. If finite accuracy measurements of an arbitrary observable of the system, made at regular time intervals from the beginning of time to the present, do not permit precise prediction of the next measurement, the system has positive metric entropy, i.e., information is gained from each measurement. Another concept recently adapted from information theory is that of algorithmic complexity, which measures the unpredictability of individual orbits. An orbit is said to have maximum algorithmic complexity provided its information content cannot be encoded in an algorithm simpler or shorter than that which simply generates a copy of the orbit. Maximum complexity implies randomness or chaos, and this concept helps to eliminate the apparent paradox of random behavior in deterministic systems.

Many connections have been established between the nonlinear dynamics of conservative systems and the dynamics of dissipative systems.

General Remarks

The developments described in the preceding sections are only a small sampling of the activities in nonlinear dynamics. Systems with instabilities leading to chaotic behavior have been emphasized. The research has been highly international, with the United States making major contributions but not dominating the field. Although physicists have added something new to the problem of understanding instabilities and the transition to weak turbulence, the majority of current research in fluid mechanics is done by the engineering community (and is described in the report of the Panel on the Physics of Plasmas and Fluids).

The impact of this research is likely to be substantial because (a) the mathematical methods of nonlinear dynamics are helping to solve many previously intractable problems in physics; (b) instabilities and chaotic motion are widespread phenomena with common features; and (c) the growing understanding of the origins of chaotic behavior and the limits of predictability contribute to the foundations of physics.

CURRENT FRONTIERS

The field of nonlinear dynamics, instabilities, and chaos is undergoing a period of rapid expansion. A considerable number of important problems merit attention.

Bifurcation Sequences

As new systems are investigated, it is likely that new distinct bifurcation sequences will be observed. Furthermore, the universality of the various bifurcation sequences still needs to be explored in some cases. For example, how universal is the breakup of a two-torus in real experiments? Much work remains to be done before the different types of transition—not only those that are mathematically possible but also those that occur physically—are classified. Are there universal phenomena associated with abrupt nonequilibrium transitions? If so, what are they? These phenomena are reminiscent of, but more complicated than, those associated with first-order phase transitions.

Little experimental work has been done on systems with two or more control parameters (codimension-two bifurcations). Since theorists have begun to classify the types of bifurcations that can occur in such systems, we anticipate that this will be a fruitful area for further experimental work.

The problem of connecting the universal phenomena near the chaotic transition with the nonuniversal behavior farther away is a challenge for the future. We do not yet know how to predict in advance what bifurcation sequence will occur in a given physical situation. At present, we can only say that if there is period doubling (for example), it must have certain asymptotic properties.

The effects of sequences of instabilities on heat, mass, and momentum transport in fluids have not been adequately studied and could have practical applications (for example, in the problem of lubrication).

Patterns

There are a large number of general questions associated with the formation of patterns in condensed-matter systems. How are patterns selected? Under what conditions, if any, is there a functional whose extrema describe selected steady states? Under what circumstances are these states sensitive to initial conditions? To boundary conditions? Conversely, how can one understand the apparent inevitability of some patterns, e.g., dendritic structures? How can one understand the selection of quantized modes (for example, in Taylor-Couette flow)? Can group-theoretical methods be useful in understanding or classifying possible patterns?

What is the role of noise in pattern selection? Can one understand pattern selection in some cases as being caused by competition involving attraction to a large number of steady states? Is the system driven from one state to another by external or dynamical noise? If so, what is the source of this noise? Nonlinear theories quantifying the strength of attraction to different attractors may eventually have to be developed.

Pattern formation is a particularly interdisciplinary problem, and physicists can benefit from interaction with metallurgists, geologists, meteorologists, biologists, and chemists. All of these groups of scientists deal with the production of intricate patterns and with the progression from patterned to chaotic behavior. Can we begin to see general rules and approaches that cut across disciplines? For example, pattern selection in hydrodynamic systems has conceptual similarities to nucleation in the physics of phase transitions. Can these similarities be used to understand these processes better? In general, the process of pattern selection is likely to become much more recognized than it has been in the past for its discriminatory power to reveal mechanisms.

Numerical Simulations

Analysis will (probably) continue to provide insight into behavior just beyond the onset of instabilities and will play an important role in understanding pattern selection. However, the difficulty of extending nonlinear stability theory beyond the second instability is clear. Accurate simulations have recently been achieved for Couette flow in the regime beyond the second instability. These simulations are difficult and require about 10^5 spectral modes. An important problem for the future is the simulation of the onset and growth of turbulence, which will require even more modes and larger computers than those now available. When properly carried out, simulations have the power to help us understand the physics of these difficult nonlinear problems.

Numerical studies of low-dimensional models will also be important in the future. Much of what we know about nonlinear dynamics comes from numerical calculations, and this will continue to be the case, especially as systems with several control parameters are investigated.

Experimental Methods

Now that the usefulness of phase-space methods has been demonstrated, we anticipate that a great deal of attention will be given to the development of reliable ways of measuring such quantities as the dimensionality, Lyapunov exponents, and metric entropy of strange attractors. One important goal of this work is to understand chaotic behavior significantly above its onset and to begin to understand the relationship between chaotic time dependence and spatial variations.

Experimenters are finding that increasingly precise methods are required to answer the most interesting questions. Laser Doppler techniques can be made more precise by using correlation techniques; digital imaging methods will make it possible to study time dependence and spatial structure simultaneously.

We anticipate that cryogenic methods will continue to be important in this field. For example, dilute solutions of ^{3}He in ^{4}He are already useful in the study of double-diffusive convection (where both concentration and temperature are important), and there is some evidence that at low temperatures these solutions may convect as classical single-component fluids with novel properties.

Transition from Weak to Fully Developed Turbulence

The onset of chaos in large hydrodynamic systems is not well understood. It is important to study the loss of spatial correlation and the evolution of time dependence with increasing Reynolds (or Rayleigh) number. Such studies will require multiple probes, digital imaging techniques, and ever bigger computers.

Theory and experiment on fluid instabilities have emphasized the onset of turbulence. There has also been much work (especially in the engineering community) on strong turbulence. However, it is not known how a weakly turbulent flow at low Reynolds number evolves into fully developed turbulence at high Reynolds number. Is the growth in the dimension of the strange attractor slow or fast? Continuous or discontinuous? How many independent degrees of freedom are required to describe fully developed turbulence (10^2 or 10^{10})? At what point does the dynamical systems approach cease to be helpful? These exciting questions will certainly be addressed in the next few years.

Eventually, it will be important to understand the asymptotic properties of real turbulence better. Are statistical averages unique at high Reynolds number? Theoreticians will have to learn how to extract information about averages from manifolds of solutions. Combinations of computer simulations and analytical boundary-layer approximations will, it is hoped, bring us closer to an understanding of turbulent fluid flow and turbulent convection in particular. Perhaps theories based on optima or bounded quantities will be fruitful.

It will be important to understand the relationship between classical turbulence and quantum turbulence (in superfluid helium). The latter is thought to consist of a tangled mass of quantized vortex lines. Since the circulation about a line is strictly quantized, perhaps quantum turbulence may ultimately be easier to understand than classical homogeneous turbulence. The resulting knowledge is likely to be of engineering importance in the cooling of superconducting devices.

Conservative Systems

Despite the deep contribution to science made by the study of integrable systems, these are, loosely speaking, no more common than integers on the line. The richness and variety of Newtonian mechanics has only begun to be sampled. The full physical significance of many-body problems that are integrable (or KAM near-integrable) and that, therefore, do not obey the laws of thermodynamics is ripe for development. The possibility that all Newtonian systems having null metric entropy are analytically solvable (although they may be ergodic and mixing) has only begun to be investigated. The full and rigorous development of the statistical mechanics of conservative and dissipative chaotic systems, including the approach to equilibrium, remains an uncompleted task. The transition from complete integrability to full chaos involves a divided phase space exhibiting chaos surrounding integrable islands of stability. Do these islands decrease in size but never disappear as the energy of the system increases? Does their presence have physical consequences for the decay of correlations, as recent numerical experiments indicate?

What is the ultimate significance of the fact that initial errors in measurement or computation grow exponentially in a chaotic system? To observe the smallest details of a chaotic orbit requires unlimited precision; to predict these details analytically requires the manipulation of infinite algorithms, which, strictly speaking, cannot even be defined. One thus begins to suspect that the study of chaos will provide an especially deep physical meaning to Gödel's theorem and the limitation it implies for human logical systems, including science. Although a final resolution to these matters lies in the future, algorithmic complexity theory has already established that most real numbers in the continuum are, humanly speaking, undefinable.

The problem of long-time prediction for chaotic systems is especially interesting. Using numerical as well as analytic arguments, numerous scientists have suggested that long-time prediction is impossible. Time complexity theory is currently being used to ask how quickly a solution algorithm can predict the future of a dynamical system, i.e., how much time is required to compute the behavior, in order to obtain insight into this question of predictability.

Finally, there is the little explored question of whether chaos remains a meaningful concept in quantum mechanics. At present, there is a growing amount of convincing evidence indicating that the quantum mechanics of finite bounded systems contains no chaotic time dependence. This leads to the deep suspicion that the quantum description of a classical chaotic system does not tend to the proper classical limit; much worse, it may be that quantum mechanics was tacitly constructed on the notion of intergrability. If so, the inclusion of chaos may require fundamental changes in the foundations of quantum mechanics.

Nonequilibrium Systems

There are a number of important fundamental questions having to do with the fact that systems undergoing instabilities are generally away from thermodynamic equilibrium. How does one characterize and classify the steady states that are obtained under constant external conditions away from thermodynamic equilibrium? (It is generally assumed that such steady states have simpler behavior than arbitrary nonequilibrium states.) Are there general techniques for classifying the spatial and temporal patterns that emerge in such systems? What are the essential similarities and differences between externally driven nonequilibrium steady states, such as cellular convection or oscillatory chemical reactions, and freely equilibrating systems such as those undergoing nucleation or spinodal decomposition (i.e., phase changes)?

There is still no really fundamental understanding of the mechanisms that control solidification patterns in, for example, snowflakes, quenched alloys, or directionally solidified eutectic mixtures. Some progress has been made in developing a stability-based theory of free dendritic growth in pure substances and dilute solutions, and it seems possible that a firm mathematical basis is being established for work along these lines. The problems of predicting periodicities of cellular solidification fronts or lamellar eutectics appear to have much in common with pattern-selection problems in Rayleigh-Bénard convection or chemical-reaction diffusion systems. Despite a long history of metallurgical investigation, the solidification problem is relatively underexplored experimentally. There is a great need for precise measurements of morphological response to varying growth conditions in simple, carefully characterized model systems.

In the case of phase transitions involving nucleation or spinodal decomposition, the major opportunity now is to use neutrons, synchrotron radiation, and other modern methods to make real-time observations of nonequilibrium processes in alloy solids, polymers, and similar materials. How do states of frozen equilibrium, like glasses and spin (magnetic) glasses, fit into the picture? Can they be understood via dynamical systems methods?

New Directions

It is important to recognize that this field is evolving rapidly and will find application to many other systems. Instabilities in semiconductors are important for device applications. The nonlinear current-voltage characteristics of a great variety of materials can lead to dynamical instabilities. Chaos is known to occur in liquid-crystal instabilities. The properties of magnetic fluids, liquid metals, and dielectric fluids are also likely to result in important nonlinear phenomena. The theoretical methods of nonlinear dynamics have already found application in models of systems with competing length scales (commensurate and incommensurate phases) in condensed-matter systems. Although systems studied in the past have been at equilibrium, nonequilibrium systems with competing periodicities are also important and are now being investigated for the first time. Finally, there are many astrophysical problems (for example, pulsating stars and geomagnetic reversals) for which the methods of nonlinear dynamics may prove useful.

There are a number of important biological applications of nonlinear dynamics, especially to neural networks. The idea that memories can be represented by attractors in a dynamical system is an appealing concept that may be capable of explaining some of the important properties of memory, such as the ability to recall a great deal of information starting with a small part of it. Another neurological application of nonlinear dynamics is the explanation of certain types of neural excitations (e.g., hallucinations) in terms of symmetry-breaking instabilities in the neural network. Research on networks is closely related to advances in computer science, where efforts are being made to apply nonlinear dynamics to the behavior of computing machines.

Phenomena in which fractal structures occur in real space rather than in phase space are being actively studied. Examples include diffusion-limited aggregation and the behavior of a two-phase fluid in a porous network. These structures have patterns on all length scales and can be understood by methods that exploit this scale invariance.

Appendixes

TABLE A.1

Connections Between Subareas of Condensed-Matter Physics and Applications of National Interest

Subareas	Information Processing	Speech and Data Communications	Energy
Semiconductors	This science underlies all aspects of hardware in computer logic, memory, and storage technologies; it is also important in energy conversion devices. It is, of course, the basis of all electronic components used in a variety of practical applications.	Semiconductor science is central to the rapidly growing field of optical communications. Lasers, light-emitting diodes, detectors, and integrated photonic-electronic devices rely exclusively on semiconductor physics and materials.	Semiconductors are the active material in solar cells and find increasing application as media for the control and switching of electric power. Semiconductor electronics pervades energy technology.
Magnetism	Magnetic phenomena form the basis of disk and type storage in computers.	Magnetic storage is important for speech and data communications.	Magnets are employed in electric generators, transformers, switches, motors, as well as for plasma confinement.
Low-temperature physics	Josephson phenomena have a potential for high-speed logic devices. Cryogenic cooling to increase charge-carrier mobility is apt to become critically important as integrated circuits continue to decrease in size.	The properties of semiconductors are more easily evaluated by performing low-temperature measurements. Some optical devices at early stages of their development have performed more reliably at low temperatures.	Superconducting magnets for magnetically confined fusion is a critical application of low-temperature physics to energy problems. Superconductors have been introduced into electric generators and motors. They are under research for power and transmission.

Nonlinear dynamics, instabilities, and chaos	Nonlinear dynamics is important in understanding the properties of adaptive networks, which may well find application in future computing machines. Furthermore, many electronic devices (Josephson junction detectors and Ge photoconductors, for example) have excess noise that is due in some cases to chaotic dynamics. In general, efforts to understand the sources of noise in condensed-matter systems are likely to have an impact on improving electronic devices and computing systems.	Solitons are beginning to play an important role in optical communications through fiber-optics technology. Much of speech is a nonlinear process.	
Defects/diffusion	Control of defects is essential in semiconductor-device processing. Device degradation mechanisms are invariably associated with atomic-scale defects and their motion through materials.	The reliability of all optical devices is limited by defects and their motion. The performance of these devices relies on the accurate control of doping and stoichiometry profiles. Diffusion of dopants over atomic-scale distances leads to a degradation of the device.	Storage batteries and fuel cells rely on ionic and mixed conductors. Defects and atomic diffusion are responsible for fatigue and corrosion, therefore for the service life of energy converters. Controlled introduction of certain defects strengthens metals.

TABLE A.1 *continued*

Subareas	Information Processing	Speech and Data Communications	Energy
Surfaces/interfaces	Epitaxy, adhesion, electrochemistry, and lubrication are all phenomena that are important to computer technology. Schottky barriers are inherently important in the operation of transistors. All these phenomena rely on surface and interface science.	All semiconductor optical devices are layered structures involving interfaces exposed to the environment. Their operation is related directly to the interface crystal and band structure. Optical fibers have several layers of glass designed to produce a chosen refractive-index profile and are protected from the environment by a polymer coating. Improvement in ohmic contacts to III-V semiconductors is urgently required.	Surfaces are modified with increasing sophistication to reduce friction and corrosion and to improve catalysts. Electric contacts and switches rely on a detailed understanding of metal surfaces.
Electronic properties	The relative stability of materials when they are in contact with each other is determined by their thermodynamic properties. This is usually obtained from measurements carried out by a computer or laboriously in a laboratory. As our ability to compute from first principles increases, we shall be able to search for new materials and their properties in a computer.	Understanding of the electronic properties of semiconductors limits the performance of devices, such as lasers or detectors, in which electronic information is transformed to optical pulses or vice versa. The band structure of materials determines their optical properties. Higher-speed electronic circuitry is required to drive photonic devices.	A fundamental understanding of electronic properties is the basis of predictive theories for most energy technologies. Electronic properties underlie such processes as conductivity, catalysis, and hydrolysis, and they also play a significant role in solar-cell technology.

...ingly important for structur-al materials and working fluids as combustion temperatures are raised for higher efficiency.

phenomena

storage devices using reversible amorphous to crystalline trans-formation devices may become practical. The glass transition temperature occurs in metals, ceramics, and a variety of sili-cate and semiconductor glasses. These materials, in one form or another, are used in electronic devices. It is important to know what microscopic phenomena determine the glass transition temperature.

...y of materials, seen as glass fibers, requires the avoid-ance of phase transitions. Liq-uid-crystal displays use a phase transition to transmit informa-tion.

Liquids

There is a possibility that liquid crystals may find extensive ap-plication as display materials.

Liquid-crystal displays are growing in importance; the development of bistable devices requires the synthesis and understanding of new liquid crystals.

The physical theory of liquids in equilibrium and of their trans-port properties forms the basis for energy-efficient design of heat engines and chemical reac-tors, including oil refineries.

TABLE A.1 *continued*

Subareas	Information Processing	Speech and Data Communications	Energy
Phonons/electron-phonon interactions	The science underlying this field is important in describing heat transport through solids or for generating new instruments such as the acoustic microscope. Surface acoustic waves are finding applications in signal-processing devices.	The recombination mechanisms in semiconductors can be radiative or nonradiative. The latter, through electron-phonon interactions, should be minimized to give maximum optical efficiency. The electron-phonon scattering mechanisms are also relevant to the problem of thermal boundary resistances, which will limit the heat extracted from both photonic and electronic devices, and hence affect their speed of operation.	The electron-phonon interaction explains the phenomenon of superconductivity and is one of the mechanisms responsible for electrical resistivity.
Polymers	These materials find widespread applications in device manufacturing where they are used as photolithographic, insulating, or substrate materials.	Polymers are extensively used as packaging, fiber coatings, and connectors and may in the future be used as single fibers for short-distance transmissions and as part of integrated optical circuits.	Polymers are widely used as electrical insulators and for energy savings through lightweight and thermal insulation.

Semiconductors	The automobile industry is projected to be one of the largest users of microprocessors outside of computers. Electronic components and computers are essential for space, air, and terrestrial transportation.	Semiconductors are the key hardware ingredient for the control, testing, and operation of spacecraft and their supporting systems.	Same as for information processing and speech and data communications, transportation, and space technology, which are all important to national security. Electronics and lasers (low-energy solid-state) have probably had the single largest impact on defense and will continue to be of prime importance in the future.
Magnetism	Permanent magnets are widely used in nearly all electric motors.	The same as for information processing and speech and data communications.	Magnetic phenomena form the basis of disk and tape storage in computers. This is an important area to DOD.
Low-temperature physics	Superconducting transmission lines may some day aid in electric car energy distribution. They may have potential in levitating trains.	Superconducting gyroscopes and Josephson junctions may allow new guidance and control possibilities for space operation.	Josephson phenomena and superconducting magnets have a potential for usage by DOD. Josephson junction computers, if made available, initially will have DOD as their biggest user.

TABLE A.1 *continued*

Subareas	Transportation	Space Technology	National Security
Nonlinear dynamics, instabilities, and chaos	There are potential applications to aerodynamics.	This field is likely to have an impact on the science of weather prediction. An improved understanding of turbulence in fluids is likely to have applications in a variety of industrial and manufacturing processes as well as applications in the field of space technology, since all these fields depend on an understanding of transport in and by turbulent fluids.	This field has potential applications, e.g., to systems and control, communications networks, nonlinear mechanical systems, and aerodynamics. Weapons systems provide many examples of instabilities.
Defects/diffusion	The formability of metals (used in, e.g., car bodies or other air frames) is strongly influenced by defect properties. Defects and diffusion are also important to semiconducting processing.	System reliability specs for space use exceed those for all other uses. Improvements in and assurance of quality here must rely on developments in fundamental understanding from physics. The physics of defects, through metallurgy, has created an impact on the strength and elastic properties of metals and ceramics.	Same as for information processing. In addition, radiation damage is a major factor in nuclear applications.
Surfaces/interfaces	Adhesion, catalysis, sensing of exhaust gas constituents, friction and wear, fuse lifetime, electrical contact integrity, and corrosion are some of the applications of surface and interface science to the transportation industry.	Interfaces are important in ion propulsion, tribology (in vacuum), fluid behavior in zero gravity. Surfaces play a significant role in solar-cell technology.	Same as for information processing.

Electronic properties	in metals and alloys, electronic properties control cohesion and binding as well as stability (fatigue and corrosion). Electronic properties are also intimately involved in surface and interface phenomena.	New transducers, radiation detectors and generators, and solar cells.	Same as for information processing, plus radiation effects on electrical materials.
Phase transitions/critical phenomena	Magnetic alloy research has a basic dependence on phase transitions as does research in steels.	Shape-memory alloys for programmed mechanical motion, amorphous materials for coatings and bulk components to achieve improvements in corrosion and tribological performance (durability and reliability of space-use components) as well as mechanical and magnetic properties.	Same as for information processing, plus some nonelectronic detectors, e.g., chemical toxin detectors. Phase transitions in actinides have a critical effect on nuclear weapons.
Liquids	Liquid crystals are being developed for display applications.	Surface tension and surface tension gradients control liquid behavior in low g. Knowledge necessary for liquid-fuel management in spaceflight.	Liquid crystals are used for a variety of displays, plus the use of fluidic devices has the potential in military and nuclear (high-radiation) applications.
Phonons/electron-phonon interactions	Heat transport through solids is of continuing interest in all engine designs.	Improved high-temperature (e.g., engine material) performance and design are tied to thermal conductivity and thermal diffusivity.	Transport properties in solids are important in defense applications.
Polymers	Composites, paints, and other plastic components use polymers.	Polymers are used as strong, light construction materials. Conducting polymers may open up new applications.	Same as for information processing, plus the use of nonelectronic materials is anticipated with growing demand.

244

TABLE A.1 *continued*

Subareas	Medical
Semiconductors	Much of medical instrumentation is solid state, including both electronics and transducers and sensors. Microfabrication techniques developed by the semiconductor industry are used to make small probes and sensors.
Magnetism	Instrumentation relies heavily on magnetic mass memory. NMR imaging techniques are becoming important for diagnosis.
Low-temperature physics	NMR imaging is becoming an active market for large-volume superconducting magnets.
Surfaces/interfaces	The design and use of biocompatible implants and sensors demands understanding of their surface chemistry and physics in their environment.

Electronic properties	There is a weak but growing connection between the understanding of the brain and electronic analogues.
Polymers	Polymers are extensively used, not only as implants but as materials with broad uses in medical equipment (e.g., storage containers, tubing, heart-lung machines).
Nonlinear dynamics, instabilities, chaos	Fibrillation of the heart is a nonlinear process.
Phase transitions	The formation of cataracts is a phase transition.
Liquids	Many different types of liquids are present in the human body.
Phonons/electron-phonon interactions	Acoustic scanning is becoming a useful diagnostic tool.

B

New Experimental Techniques

Important experimental methods or techniques have come into being in the last decade or, if discovered earlier, have been extensively developed and utilized in recent times. In the main their descriptions have been interwoven with the discussion given in the text and in the following three appendixes.

In Table B.1 we present the utilizations of various new experimental techniques to the individual subdiscipline areas of condensed-matter physics. Most represent new horizons in higher frequencies, magnetic fields, or pressures, while others provide new ways of probing or analyzing surface structures. Undoubtedly, the advances that were achieved in condensed-matter physics in the past decade have been intimately tied to the development and extensive utilization in varying degrees of the techniques given in the table. For example, all the new discoveries in the superfluid properties of liquid ^{3}He are a direct result of the existence of the dilution refrigerator (Chapter 8). The observation of the quantized and fractionally quantized Hall effect was made possible by the development of high magnetic fields (Appendix E). Numerous examples of the utilization of the various surface-science probes (e.g., the scanning tunneling microscope, atom/surface scattering, low-energy electron diffraction) are to be found in Chapter 7, without which our understanding of the nature and structure of the surface would be considerably less sophisticated. The reader is referred to individual chapters and appendixes for detailed descriptions of these techniques and their applications.

246

TABLE B.1 New Experimental Techniques and Their Principal Applications to Condensed-Matter Physics

Technique	Electronic Properties	Structure and Vibrational Properties	Critical Phenomena	Magnetism	Semiconductors	Defects and Diffusion	Surfaces and Interfaces	Low-Temperature Physics	Liquids and Liquid Crystals	Polymers	Nonlinear Dynamics, Instabilities, and Chaos
Synchrotron radiation	X	X		X	X	X	X			X	
Laser techniques	X	X		X	X	X	X		X	X	X
Molecular-beam epitaxy	X	X		X			X				
Diamond anvil cell	X	X									
Surface analysis techniques	X		X				X				
High magnetic fields	X		X	X				X			
Dilution refrigerators	X		X	X				X			
Neutron scattering	X	X	X	X		X	X	X	X	X	X

C

New Materials

INTRODUCTION

The creation of new materials, always a driving force toward advances in condensed-matter physics, has in the past decade become an even more important, perhaps dominant, factor as our understanding of general physical phenomena (e.g., conductivity, magnetism, and superconductivity) has progressed to the point where intellectual challenge is now provided by departures in specific materials from the general behavior, rather than in this behavior of the class of materials as a whole. In many cases these departures represent the limiting behavior that also makes the phenomenon of technological value, for example, the highest stored energy density or lowest hysteresis loss in magnetic materials, the highest mobility in semiconductors, or the highest critical temperatures, magnetic fields, and currents in superconductors. In this appendix we first identify the materials that have had a major impact on condensed-matter physics in the past decade, second understand how they were discovered, and third recommend how the chances of such discoveries can be maximized in the future. Given the record of the past, such discoveries are essential if the vitality of condensed-matter physics is to be assured.

We define as new materials those that have led to the observation of new physical phenomena. Thus, new processing techniques, unless

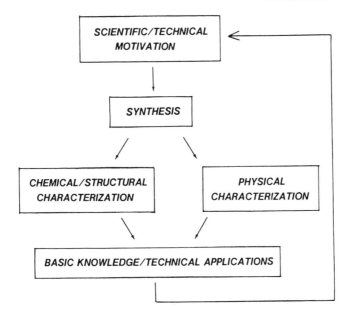

FIGURE C.1   The synthesis loop.

they have resulted in the synthesis of materials with unusual properties, will at most be only briefly mentioned.

An underlying theme of this section is the importance of the synthesis loop (Figure C.1). This interrelationship between synthesis, the initial characterization of materials, and the evaluation of these materials, from which may come (unpredictably) either basic knowledge or technical applications, must be closed by feedback of these results to the individuals carrying out novel and creative synthesis. The motivation for completing this synthesis loop can be scientific or technological.

NEW MATERIALS IN THE LAST DECADE

One- and Two-Dimensional Transition-Metal Chalcogenides

This family of materials has created great activity as a result of the observation, in 1974, of charge-density waves in them (see Chapter 1). This has led to questions of how charge-density waves move and of their pinning and to the prediction and observation of boundaries between commensurate and incommensurate regions of the wave. The charge-density wave has been observed directly by atomic-resolution electron microscopy (see Figure 1.4).

Materials with Open-Crystal Structures

Although intercalated graphite is an old material, there has been considerable interest in the staging of the intercalants, the steps whereby spaces between the two-dimensional carbon layers become filled.

In solid electrolytes the crystal structure is open enough that ion transport can occur through the material. Activity in this field is high, driven by the need for massive batteries for load leveling and mobile power sources. Some of the chalcogenides are also fast ion conductors when a third element (such as Li) is added. These materials are both electronic and ionic conductors and could be used as electrodes (not electrolytes) in high-energy density batteries.

Zeolites can be used to support extremely small metallic particles or can be filled with metal, which results in an interesting network of filaments whose diameters are ~10 Å.

The Magnetic Superconductors

The history of this field is a fascinating one with respect to the question of whether superconductivity and magnetism could simultaneously exist in the same material (see Chapter 8). In 1967 and 1971, members of a new family of ternary materials, now known as the Chevrel phases, were shown to exhibit superconductivity. Another new ternary compound, $ErRh_4B_4$, was first shown to be superconducting at 8.6 K but then to return to the normal state with the onset of ferromagnetism at ~1 K. Chevrel phases were shown to exhibit superconductivity, which was then destroyed by ferromagnetism at a lower temperature.

There followed exceptionally productive examples of the synthesis loop, where an extremely close coupling between basic scientific evaluation, theoretical input, and synthesis led to rapid progress and understanding of how superconductivity, ferromagnetism, and antiferromagnetism can coexist.

Organic Conductors

A major advance in the past decade was the discovery that organic materials exhibit not only high metallic conductivity but also superconductivity (see Chapter 8). The materials of interest fall into a few groups. The so-called Bechgaard salts have remarkable transport properties (e.g., an unusually strong decrease in resistivity with decreasing temperature, whose rate is strongly affected by an applied magnetic field). Superconductivity was observed initially in one of them at ambient pressure and also in the organic compound polyacetylene, but under pressure. Although technically of considerable interest, the fascination of this material lies in its physical properties, especially the nature of the charge carriers (solitons; see discussion in Chapters 2 and 10) produced in it by doping.

New Superconductors

Just before his untimely death B. T. Matthias discovered $Ba(Pb_xBi_{1-x})O_3$, which has the highest transition temperature (T_c) for compounds not containing a transition metal. We now know that this T_c is due to extremely strong electron-phonon coupling (the well-known mechanism of superconductivity) in a material with a low density of electron states rather than to an unusual coupling mechanism. More recently heavy-fermion superconductors have been discovered, having relatively low T_c but anomalously large values of the electron mass, examples of which are $CeCu_2Si_2$ and UBe_{13}.

Glasses

This broad field encompasses the structure of glasses, transport and the metal-insulator transition, thermal properties at low temperatures, spin glasses, amorphous semiconductors, solitons in high-purity optical fibers, and the effects of disorder and localization on superconductivity. The synthesis methods used in the creation of glasses include not only rapid quenching of bulk materials but also the quenching of films onto low-temperature substrates, first used to produce an amorphous material in 1954.

Artificially Structured Materials

The most remarkable advance in the past decade was the creation of what are known as artificially structured materials (see Chapter 1). The classic example of this synthesis technique is the growth of semiconductors by molecular-beam epitaxy, which defines not only their layers but also the specific location of their constituents, dopants, and charge carriers. In addition to the enormous technological importance of this synthesis technique, it has led directly to new physical phenomena, including the quantized Hall effect and the fractionally quantized Hall effect (details are given in Chapter 1).

IMPACT OF NEW SYNTHESIS TECHNIQUES ON CONDENSED-MATTER PHYSICS

Thin Metallic Films

A rapidly expanding field is that devoted to the conductivity of materials at low temperatures, especially those that undergo a transition from metallic to insulating behavior because of disorder or reduction in their size. Interest in this field was stimulated by specific predictions of the maximum resistivity that can be reached in metals and made possible by advances in low-temperature techniques (Chapter 8), by improvements in thin-film deposition methods, and by the development of the photolithographic and processing techniques that allow patterning of the deposited films to dimensions $\ll 1$ μm. Popular materials for these studies include the two-dimensional electron gas in MOSFETs, in which the carrier density is tuned by the gate voltage, and codeposited mixtures of metals and insulators, where the carrier density is varied by changing the insulator content. Such studies have given new insight into the nature of electronic transport, elastic- and inelastic-scattering processes, the localization of electrons in disordered metals and the interactions between them, and even the phase coherence of the electron wave function in nonsuperconducting metals.

Epitaxial Materials

The remarkable success of epitaxial growth in the synthesis of semiconductor materials has created new interest in many other systems, for example, metal on semiconductor, metal on metal, semiconductor on insulator (or vice versa), and insulator on insulator.

A heavily studied class of metal-on-semiconductor materials is the family of transition-metal silicides on silicon, which have technological importance as conductors in VLSI circuits. One can grow such silicide films as perfect single-crystal films by a novel technique that uses a template for epitaxial growth.

The growth of an epitaxial insulator on semiconductors again has technological implications, especially in the case of semiconductors that do not naturally grow an insulating oxide with the quality of silicon oxide. Interestingly, the degree of lattice match between insulator and semiconductor does not seem to determine the quality of epitaxial growth. An ultimate aim of this research is the ability to grow repeated insulator and semiconductor layers, which then might be processed into three-dimensional circuits. The properties of such thin semiconductor layers, especially when patterned to $\ll 1$ μm in the lateral dimension, will be of interest in determining the physical limits on the sizes of electronic components.

The deposition of metals on insulators has begun to benefit from advances in vacuum and surface-science techniques, and the growth of such layers is now taking place in ultrahigh vacuum, with *in situ* analysis of the growing film. The ability to control substrate temperature and to deposit simultaneously or sequentially from multiple sources has led to the development of extremely versatile materials synthesis systems and may, to some extent, replace the traditional synthesis of bulk samples. Control of the substrate temperature gives a new dimension to this synthesis method, allowing the easy growth of amorphous and metastable phases. As an illustration, the highest superconducting transition temperature, that for Nb_3Ge ($T_c = 23$ K), was achieved only by vacuum or chemical vapor deposition of Nb and Ge onto a carefully temperature-controlled substrate.

Control of metastability has also been achieved by metal-on-metal epitaxy. Studies of metal-metal interfaces using ion channeling indicate that interfacial strain can induce lattice match between two metals (or semiconductors) over moderately long growth distances. This results in an alternative method for the synthesis of metastable Nb_3Ge, namely by epitaxy with stable Nb_3Ir, which has the same lattice constant. Nb_3Al was also stabilized beyond the bulk-phase boundary by self-epitaxial growth, namely by continuously changing the Nb/Al ratio away from the stable composition during growth of the film.

Modification of the physical properties of a surface, and thus indirectly of the bulk material, has also been achieved by metal-on-metal overlayers, although epitaxy is not essential. For example, the absorption of hydrogen by bulk niobium was drastically increased by the deposition of a few monolayers of palladium on its surface. The inactivity of Au for catalyzing the rate of cyclohexane dehydrogenation to benzene was increased above that of Pt by the presence of two monolayers of Pt on the Au surface. The activity of bulk Pt for the same reaction was increased fourfold by the addition of a monolayer of Au.

Metallic Superlattices

The logical extension of work on metal epitaxy is to repeat the process many times, thus forming a metallic superlattice. The high level of recent activity in this field primarily addresses questions of structure and how the behavior of these multilayer materials is largely determined by the interfaces, which become essentially the whole of the material as the superlattice period is reduced. In only one system, Nb/Ta with perfect lattice matching between Nb and Ta, do films have an electronic mean free path that is appreciably longer than the period. In others, the mean free path is limited by scattering at the interfaces. In Nb/Ta, the phonon distribution function, measured by tunneling, clearly shows the alloying at the interfaces, as does x-ray scattering.

Two surprising observations have come from these studies to date. One is the anomalous change in the elastic properties of some superlattices as the period is reduced to about 30 Å. The other is the ease with which it appears possible to induce new lattice structures by epitaxy and interfacial strain. Thus Zr can be stabilized in the bcc phase in Nb/Zr superlattices, as can Co in Co/Cr. Both Mo/V and Nb/Al have been shown to grow as strained superlattices, in exactly the same way as the strained semiconductor superlattice described in Chapter 5.

Superlattices of metal with insulator, for example, Nb/Ge, constitute ideal systems for the study of dimensional effects in metals, in this case the crossover from two-dimensional to three-dimensional superconductivity as the Ge thickness is reduced. It is clear that such research will continue with the goal of achieving single-crystal metal/insulator (or semiconductor) superlattices.

Materials Modification

New techniques for the modification of materials have emerged that both challenge current theories of crystal growth and have technological value. For example, ion beam/solid interactions are being used to dope and amorphize semiconductors and to produce new metallic alloys by implantation and ion mixing, while the interaction of ion beams with organic films produces conducting layers and has applications to lithography.

Laser processing of semiconductors can result in single-crystal growth and in anomalously high levels of incorporation of dopants. Laser annealing of semiconductors promises to improve the quality of devices and save billions of dollars for the electronics industry. Laser alloying is also possible, creating alloy layers of extended solutions that are otherwise difficult to make. Lasers can be used to modify the structure of surface layers or films in an important manner. Depending on the laser pulse width used in heating, laser annealing or quenching can yield either amorphous or crystalline films of various materials. For surface modification, laser-induced photochemical deposition, chemical etching, and electroplating have been demonstrated to provide order-of-magnitude improvement in speed over conventional methods.

The fabrication of regular arrays and random distributions of metal particles in the 100-1000 Å size range by deposition techniques, laser processing, and microlithography has helped to elucidate the phenomenon of surface-enhanced Raman scattering. This is important as a probe of surface structure and adsorption and has improved our understanding of optical processes at surfaces.

Explosive techniques are beginning to be used for the preparation of novel materials. For example, the Al5 superconductor Nb_3Si has been prepared with its highest transition temperature by this method, which also induced anomalous properties in the semiconductor CdS.

Filamentary Materials

By drawing from an ingot containing two components (say wires of Pt embedded in a Cu cylinder) it is possible to produce a fine wire in which the Pt exists as filaments <1000 Å in diameter. For example, drawing a Cu ingot containing a dispersion of Nb particles results in a multifilamentary wire that can be reacted to form superconducting Nb_3Sn filaments. The interest to date in these multifilamentary materials has centered largely on their mechanical properties, but it has also been shown that by etching away the Cu matrix, fine single filaments can be produced that have unusual physical properties: high electrical resistivity and high strength, for example. Given the interest, mentioned earlier, in systems with dimensions <1 μm, this alternative method of making free-standing material with no strain at a substrate should be explored further.

Metal Clusters

The production of small clusters of a material by its evaporation into an inert gas, so that the particles agglomerate before condensing on a substrate, has been used for some time to produce three-dimensional samples of very small size, say <100 Å in diameter. Recent interest in the physical properties of these materials is beginning to address the question: How large is a particle before it assumes the bulk properties of the material? A new approach to studies of metallic clusters has been to measure their mass distributions by time-of-flight techniques. There is clearly indication that certain (magic) numbers of atoms in a cluster are most stable. Novel sources of such clusters have been developed, including one in which a pulsed laser is used to vaporize a metal in a supersonic expansion nozzle.

MAJOR CONCERNS

From the discussion above, it is clear that materials that have received the greatest attention in the last decade can be divided roughly equally into bulk materials and artificially structured materials. The major concerns of the 1980s that can be identified from what has been said in this appendix are different with respect to these two classes of materials. In the field of artificially structured materials the central and crucial role of synthesis to the success of any program of research has emerged in a natural way. Either one person is responsible for all aspects of synthesis, characterization, and evaluation or a closely knit group has emerged that is built around state-of-the-art facilities for synthesis, say of semiconductors by molecular-beam epitaxy (MBE) or metallic films by codeposition, or for processing to submicrometer dimensions. In the case of processing, state of the art is an overstatement except in a few laboratories: in the majority the inventiveness of the individual scientist has to overcome the limitations of antiquated equipment. Thus, many novel methods for producing submicrometer structures have emerged. The main concern in this area is the extremely high cost of initiating research. For example, an MBE system costs about $500,000, a high-vacuum metals codeposition system with no surface characterization tools in place costs more than $300,000, and an electron-beam pattern-writing machine costs $1 million (although electron microscopes can be modified for this purpose at lower cost).

This natural role of synthesis should be contrasted with the situation in research into new bulk materials, where synthesis is fragmented if it exists at all.

PROJECTIONS FOR THE FUTURE

Judging from the directions taken by condensed-matter physicists in the field of new materials in the past decade, some projections can be made into the next decade.

First, driven by technology and scientific challenge, great progress will be made in artificially structured materials on length scales that will be reduced below 100 Å. The epitaxial growth of metals, insulators, and semiconductors in all combinations and as superlattices, and of clusters, will lead to new phenomena and new devices.

Second, in studies of bulk materials, physicists will continue to extend their interest away from perfect single crystals of simple materials to materials with complex structures and large unit cells, especially structures with internal clusters or internal channels where properties can be modified by intercalation or ion insertion.

Third, the U.S. condensed-matter community in general will continue to contribute to materials evaluation of bulk materials rather than to their synthesis.

Fourth, the evidence of the past decade is that the discovery of new materials will lead to the discovery of new physical phenomena. In some way, funding of science in the United States must be structured to encourage this to happen.

D

Laser Spectroscopy of Condensed Matter

INTRODUCTION

The advent of lasers has greatly facilitated research in materials science. That a laser beam has the characteristics of high intensity, strong directionality, and extreme monochromaticity, and can appear in ultrashort pulse form, makes it a unique tool for materials studies. In the past two decades, a large number of laser techniques have been invented for the investigation of matter in all its phases. They have opened the possibility of research in many hitherto unexplored areas of materials science.

Generally speaking, lasers can be applied to two types of problem: (1) to probe a material and (2) to modify or process a material. The past decade has witnessed great advances along both lines. For example, laser light scattering has become a conventional technique for studying excitations in condensed matter; nonlinear optical spectroscopy allows the study of forbidden transitions in a medium and the study of homogeneous broadening of spectral lines as narrow as ~ 1 kHz; transient optical spectroscopy can probe dynamic properties of a medium on a time scale as short as a subpicosecond (i.e., $<10^{-12}$ second); optical mixing is useful for monitoring and studying molecular adsorbates on surfaces; and laser heating is promising as a new method for annealing crystalline films or for growing various types of amorphous, crystalline, and alloy layers. As laser techniques become increasingly more sophisticated, we anticipate in the coming years an exciting period for laser-related materials research, from both the practical and scientific points of view.

ACCOMPLISHMENTS OF THE PAST DECADE

Spectroscopic study is essential for the microscopic characterization of a medium. Laser spectroscopy has brought new life to optical spectroscopy of condensed matter. In this section we survey the accomplishments of the past decade.

Nonlinear Optical Spectroscopy

This type of spectroscopy flourishes only because tunable lasers have become easily available. It comes in many different forms, depending on the nonlinear optical process involved. A few of them are considered here.

TWO-PHOTON SPECTROSCOPY

Two-photon spectroscopy is commonly used to study transitions between states of the same parity. In research on exciton polaritons in a semiconductor, for example, since two-photon spectroscopy is unaffected by the reststrahlen band it can yield detailed information about the damping and dispersion curve of the exciton polaritons. The technique has also been applied to the study of excitonic molecules in solids, which is a subject of immense theoretical interest. More recently, two-photon spectroscopy has been used to measure the $(4f)^n \rightarrow (4f)^n$ transitions of rare-earth ions in solids not observable in one-photon absorption owing to the presence of the $(4f)^n \rightarrow (4f)^{n-1}$ transitions. This is most interesting because through such an investigation one can expect a much better understanding of rare-earth ions and their interaction with the lattice in a solid.

HOLE BURNING IN INHOMOGENEOUSLY BROADENED SPECTRA

A laser beam can be intense enough to saturate a transition in a medium. If the laser linewidth is much narrower than the inhomogeneous broadening of the line then, with the laser frequency fixed, only a small group of ions or molecules can be resonantly excited by the laser. By saturating the transition of this group of ions or molecules, a hole is created in the inhomogeneously broadened spectrum. The hole width is generally limited by homogeneous broadening, assuming that the laser linewidth is negligible. This hole-burning effect makes high-resolution spectroscopy of ions or molecules in condensed matter a reality. Both absorption or fluorescence can be employed to probe the holes. Lines as narrow as a few megahertz have been observed. This saturation-spectroscopy technique can be used to study hyperfine and superhyperfine interactions of rare-earth ions in solids and fine structure of organic molecules in solid matrices. At sufficiently low temperatures, the laser-induced holes can last almost indefinitely. As many as $\sim 10^5$ holes can be burned on an inhomogeneously broadened line 10 cm^{-1} wide. They can therefore be used for making optical memory devices with high densities of data bits. Such a possibility is currently being pursued vigorously by several industrial laboratories.

OTHER NONLINEAR SPECTROSCOPY TECHNIQUES

Four-wave mixing and coherent Raman spectroscopy allow us to study excitations in both the ultraviolet and the infrared ranges. Because their sensitivity is high, they can be employed to detect impurities in condensed matter. Their spectral resolution is limited only by the laser linewidth. Because the output is coherent and directional, spatial filtering can be adopted to suppress the luminescence background. Thus, these techniques can be used to study excitations that are normally masked by luminescence. Applications of these techniques to condensed-matter physics have attracted only limited attention in the past; but more recently, with the advance of laser technology, they have begun to receive increasing recognition in the community.

Transient Optical Spectroscopy

Transient coherent phenomena arising from the resonant interaction of radiation with matter are among the most fascinating topics in condensed-matter physics. They were studied extensively in magnetic resonance before the laser era. With lasers, extension of these studies to optical transitions becomes possible. The past decade has seen increasing activity in this area of research. With the use of photon echo and optical nutation techniques, for example, homogeneous linewidths of optical transitions of rare-earth ions in crystals as narrow as ~ 1 kHz have been measured. Thus, detailed information about spin-spin interactions between the ions and surrounding atoms can be obtained.

Ultrafast Laser Spectroscopy

PICOSECOND LASER SPECTROSCOPY

In 1966, laser scientists discovered how to use the technique of mode locking to produce optical pulses as short as ~5 ps in duration. Stimulated by progress in laser technology, the past decade has seen dramatic improvements in picosecond instrumentation and techniques, with the result that picosecond laser spectroscopy is an area experiencing major advances at present. Several laser manufacturers now sell reliable cw mode-locked dye lasers delivering ~3-picosecond-long pulses at ~100-MHz repetition rates. These pulses may be selectively amplified to energies in the millijoule regime (i.e., having peak powers of ~10^8 watts), and pulse length measurements, including delay times, are routinely carried out by the technique of autocorrelation by nonlinear mixing.

In general, by exciting a medium with a picosecond pulse, followed by probing with another picosecond pulse, the dynamic properties of the medium can be studied on a picosecond time scale. This is exciting since it offers the opportunity to measure directly, for the first time, the carrier relaxation time, the excitation lifetime, the phonon relaxation, and other properties in a condensed medium. It opens a new, important area of research that seldom has been explored in the past.

As an example, time-resolved photoemission spectroscopy (PES) of semiconductor, metal, and insulator surfaces yields important data on the transient behavior of selectively excited carriers. Using a picosecond laser pulse to excite a narrow energy distribution of electrons, a delayed picosecond pulse has been used to probe the relaxed distribution by PES. By energy analyzing the photoemitted electrons as a function of time delay, fundamental information about the energy relaxation processes affecting the electron distribution was obtained. Furthermore, by using circularly polarized laser light and studying the spin polarization of the photoemitted electrons, phase-destroying processes may be studied. Thus momentum relaxation processes may be distinguished from energy relaxation processes.

In a new approach to high-speed electronic instrumentation, known as picosecond optical electronics, a short laser pulse illuminates a high-speed photoconductor, thereby producing a fast switch for an electrical signal. By combining two of these switches with a controllable time delay, a sampling system capable of time resolution of better than 2 picoseconds has been demonstrated. Studies of propagation delays between the drain and gate signals of GaAs ferroelectric transistors have been carried out. The response time of a high-speed silicon photodiode has been measured. Electron transport in materials with particular types of electronic defects has been characterized with picosecond time-resolved photocurrent spectroscopy. The extension of optical electronics to the subpicosecond regime calls for ways of overcoming dispersive and capacitive effects in electronics components. Efforts are being devoted to finding better ways of rectifying the light pulses to produce short dc pulses (such pulses really look like microwave radiation since they contain frequencies in the terahertz range).

FEMTOSECOND LASER SPECTROSCOPY

In several research laboratories techniques were developed recently to produce laser pulses as short as 70 femtoseconds (1 femtosecond = 10^{-15} second). A method of pulse compression has recently been developed that has resulted in the realization of pulses only 16 femtoseconds long. One of the results that has been achieved by the use of such ultrashort laser pulses is described in what follows.

An area of great interest over the past several years has been laser processing and laser annealing of materials. While many different facets of this field have been explored by studying the results of laser irradiation of materials, little has been done to time-resolve the actual annealing process. Recently, such studies have been carried out with femtosecond pulses, and the results support a model where the light is absorbed and first creates electron-hole pairs, after which the irradiated surface is converted from a crystalline structure within an electron-hole plasma to a molten state. By such studies, one can determine the dynamics of the energy absorption process responsible for laser processing. This is an exciting new area of materials science.

SOME DIRECTIONS FOR FUTURE RESEARCH

Now that laser technology has become more mature, one can anticipate rapid growth in several areas of laser applications to condensed-matter physics in the coming years.

The increased use of two-photon spectroscopy for the study of forbidden transitions of rare-earth ions in solids is anticipated.

In studies of materials properties, high-resolution laser spectroscopy and nonlinear optical spectroscopy are expected to become common laboratory techniques.

The future of transient optical spectroscopy for the study of optical transitions of rare-earth ions in crystals is particularly bright because these techniques can be extended to the study of transitions between excited states. Many of the sophisticated techniques developed earlier in magnetic resonance are yet to be transplanted to this field.

Another area of rapid growth in the laser spectroscopy of condensed matter is laser studies of surfaces. Both laser perturbation and laser probing of surfaces are exciting new areas of research that have hardly been explored.

It is clear that we have entered a new era in making measurements on subpicosecond time scales. Laser-based techniques will allow such measurements to be made routinely in physics, chemistry, biology, materials science, and device studies. This area of research is now known as femtosecond science. Some of the interesting subjects for study in this area are listed below.

By using a femtosecond laser pulse to create photoemitted electrons with a small energy spread and by accelerating these electrons across a large potential, a pulse of electrons is made available for time-resolved electron-diffraction studies. For the first time, such structural changes as melting and structural phase transitions may be studied on the femtosecond scale.

The response of crystalline solids to a short electrical pulse, i.e., the electric field of the laser pulse, may be studied by femtosecond spectroscopy. For example, the Franz-Keldysh effect (the change of the band gap of a semiconductor in an electric field) can be studied in a time-resolved fashion.

When a femtosecond light pulse causes photoemission, the electron pulse and the light pulse are synchronized. This makes it possible to carry out studies where one pulse creates an excitation and the other pulse probes this excitation. Thus the electron pulse may create a change in matter than can be monitored optically by the light pulse. It might be possible to create or disturb a solid-state plasma using the short electron pulse and then probe it by time-resolved reflectivity measurements.

In the area of femtosecond-pulse technology, research is aimed not only at producing still shorter pulses but also at developing techniques to extend the wavelength region of coherent-light generation toward the ultraviolet and the infrared regions.

On the side of laser technology we can also expect to see advances in several areas. A severe limitation in the progress of high-power laser technology comes from optical breakdown. This is a subject of great practical importance and has been pursued vigorously in the past two decades. From the scientific point of view, this is also a subject of great interest. Optical breakdown in solids is a highly nonlinear process. How a laser beam excites the carriers, generates the plasmas, and eventually shatters or melts a solid is not a trivial matter to understand. Breakdown mechanisms for excitations with different laser frequencies and pulse widths could be different. Solution of the problem requires the joint efforts of theorists and experimentalists. Progress in this area is being made step by step. It is hoped that in the next decade, optical materials of much improved quality will be produced as a result of the continuing research effort.

Another important area of materials research related to laser technology is the search for better nonlinear optical crystals. Such crystals are essential in the extension of coherent light sources to the ultraviolet and infrared regions. Recent theoretical calculations are fairly successful in predicting nonlinearities of certain crystals. Organic and inorganic crystals of large nonlinearities have actually been grown following the theoretical hints. This is encouraging. It is likely that new and better nonlinear optical crystals suitable for efficient frequency conversion over a broad range and for use in nonlinear optical devices for data processing will emerge in the near future.

Optical fibers have grown, in the past decade, into an important branch of the optical industry. They are the key element for future optical communications and data processing. An exciting development in the research into fiber materials has taken place within the past year. It is found that crystals can also be pulled into thin optical fiber form. Thus, fiber lasers and fibers for optical frequency conversion may soon appear in laboratories. Their possible applications are numerous, limited only by imagination. They are likely to revolutionize science and technology in many disciplines. The prospect of this field is truly great. Its progress in the next 10 years will be interesting to follow.

E

National Facilities

INTRODUCTION

Condensed-matter studies have traditionally been small-scale physics, performed at laboratories located in the scientist's home institution, be it a university, governmental, or industrial laboratory. However, with the advent of neutron scattering in the 1950s and, more recently, synchrotron radiation, as important probes of solids and liquids, a significant fraction of condensed-matter physics research requires the use of extremely costly facilities that are available only at national laboratories. In addition, special environments (e.g., high magnetic fields) required to understand the condensed states of matter are again beyond the financial resources of individual institutions. For both reasons the national facilities at either national laboratories or particular universities have come to play a special and important role in condensed-matter physics research.

The two most substantial items in the national facilities budget are for sources of neutron beams and of synchrotron radiation. Table E.1 explores how these two techniques are utilized in different and complementary ways in condensed-matter physics studies.

Table E.1 Fundamental Characteristics of Neutrons and X-Ray Beams

Neutrons	X Rays
1. Low-energy particles (1-500 meV)	1. High-energy particles (1-30 keV)
(a) Well suited to inelastic scattering of thermally generated excitations. Poorly suited for study of higher energy levels.	(a) Well suited to probe energetic electronic energy levels. Poorly suited to study of low-lying thermal excitations.
(b) Generally little radiation damage. Particularly important for organic and biological samples.	(b) Intense beams can produce radiation-induced defects in samples under study.
(c) Generally highly transparent to sample and container. Measurements in extreme sample environment (e.g., temperature, pressure) relatively easy. Measurements sample bulk properties of material under study.	(c) Matter relatively opaque to x rays. Extreme sample environments are more difficult to produce. Measurements are surface sensitive (depth of 50-5000 Å). Absorption can be used to advantage in EXAFS measurements.
2. Short-range nuclear interaction	2. Long-range electromagnetic interaction.
(a) Measurements less susceptible to complicating multiple scattering effects. Scattering efficiency is low. Large samples required for inelastic scattering studies.	(a) Scattering efficiency is high; small samples can be studied effectively.
(b) Scattering cross section varies irregularly from nucleus to nucleus. Useful to determine position and motion of light atoms, particularly hydrogen. Specific key atoms can be labeled by isotopic substitution without changing their chemical function.	(b) Scattering power varies smoothly from light to heavy elements, making light elements more difficult to study and nearby elements in the periodic table hard to distinguish.
3. Neutron carries magnetic moment and thus drives magnetic response of materials. Uniquely suited to study of arrangement of magnetic atoms and magnetic fluctuations of materials.	3. X ray carries oscillating electric field and drives the electric response of materials. This is the most universal and richest mode of excitation of all condensed matter.
4. Low brightness sources. Neutrons emitted randomly in all directions.	4. High brightness sources. Unlike conventional x-ray tubes and neutron sources, synchrotron light sources are highly directional.
(a) All inelastic-scattering experiments are compromised by low counting rates. Long counting times, poor angular resolution, and large samples are required.	(a) High angular resolution can be attained with little loss of intensity.
(b) Coherence-enchancing devices are impossible in principle, but brightness can be conserved with neutron guide tubes.	(b) Coherence-enhancing devices (wigglers, undulators) increase the brightness even further.

SYNCHROTRON RADIATION RESEARCH

Introduction

Synchrotron radiation is electromagnetic radiation emitted from particle accelerators by charged particles (usually electrons) with large energy in the range from hundreds of MeV to 10 GeV or more. With suitable instrumentation one can select a narrow bandpass of monochromatic radiation and explore a range from the visible (2-eV photon energy) to the hard x-ray region (50,000 eV or more). The current widespread interest in synchrotron radiation is illustrated by the increase in U.S. users from 50-60 in 1973 to 550-650 in 1983.

Synchrotron radiation has a number of characteristics desirable for research, e.g., high brightness,* wide tunability, strong collimation, linear polarization, stability, and the fact that the radiation occurs in 0.1- to 1-nanosecond pulses. Synchrotron sources provide intense radiation at wavelengths for which laser sources are either unavailable or not yet tunable (photon energies above 50 eV or wavelengths shorter than 250 Å). Synchrotron radiation is essentially incoherent but can be made to be partially coherent by the use of devices known as undulators that provide huge enhancements of brightness (by a factor of 1000-10,000), at certain specific wavelengths, over the already bright conventional bending magnet sources of synchrotron radiation. Similar devices, called wigglers, provide correspondingly large enhancements of the intensity of synchrotron radiation at specific wavelengths.

Summary of Present Synchrotron Facilities in the United States

We briefly describe the current status of the five synchrotron laboratories in the United States with the approximate numbers and characteristics of their user communities along with the evolutionary nature of research currently being carried out.

Cornell High Energy Synchrotron Source (CHESS)

CHESS is the parasitic synchrotron laboratory that is part of the particle-physics operation at the 4-8 GeV Cornell Electron Storage

---

* Spectral brightness is defined as the number of photons, within a narrow energy bandpass, emitted per unit area of the source and per unit solid angle. Radiation intensity is the integral of spectral brightness over the source area. Some experiments use only high intensity, while others require high brightness as well.

Ring (CESR). CHESS has supplied synchrotron radiation for about 2 1/2 years and is funded by the National Science Foundation (NSF). Because of the large circumference of CESR, single bunch operation gives CHESS a particularly useful time structure, which has been used to study x-ray diffraction on a nanosecond time scale. Three primary beam lines supply hard x rays ($h\omega > 3$ keV) to six experimental stations for general use. Three stations scan energy continuously in the ranges 3-20, 3-35, and 3-50 keV, respectively, and two other lines provide focused x-ray beams. CHESS has ~75-85 active proposals with one third from Cornell faculty and staff, one third from other universities, and one third from industrial and national laboratories.

National Synchrotron Light Source (NSLS)

The NSLS at Brookhaven National Laboratory is a Department of Energy (DOE)-funded facility with a 750-MeV electron storage ring for ultraviolet (UV) and soft-x-ray research and a 2.5-GeV storage ring for x-ray research. Each ring is designed for high brightness from conventional bending magnets, and each is 100 percent dedicated to photon production. The UV ring has recently completed its first year of operation, whereas the x-ray ring is at an early operational stage and using photons for alignment of instrumentation. There are 16 ports on the UV ring and 28 ports on the x-ray ring; each port can accommodate several experimental stations. Twenty-two experimental stations are in operation or under construction on the UV ring, and 34 stations are under construction on the x-ray ring. About 60-70 percent of the beam lines are being funded and constructed by participating research teams (PRTs) from a variety of university, industrial, and national laboratory groups. The PRTs receive up to 75 percent of the beam time for their investment in equipment and support, with the other 25 percent as well as ~100 percent of time on NSLS user beam lines available to the general scientific community on a proposal basis. At present there are 34 PRTs with a total membership of nearly 200 scientists. A free-electron laser on the UV ring has been installed, and several undulators and wigglers are under development for installation in the x-ray ring in the near future.

Stanford Synchrotron Radiation Laboratory (SSRL)

SSRL is a DOE-funded facility at the Stanford Linear Accelerator Center that utilizes the 1.5-4 GeV storage ring, SPEAR. Fifty percent of its time is dedicated to photon production, and the remainder is available for parasitic use of synchrotron radiation during particle-physics experiments. At present, there are 18 experimental stations on 7 ports, of which 3 are conventional bending magnet ports and 2 are either 8-pole wigglers or 30-period undulators. Two new beam lines are being developed in joint SSRL and PRT collaborations: one has a 54-pole wiggler, and the other has interchangeable undulators to optimize different portions of the spectral region 10-1000 eV. One of the remaining straight sections is committed to development of an in-vacuum undulator capable of producing 8-keV x rays, and another section is for an additional multipole wiggler to be developed in collaboration with outside PRTs. Almost all of the available time is assigned to the general scientific user community on a proposal basis. At present there are 150-200 active proposals from a user community of approximately 200 scientists.

Synchrotron Radiation Center (SRC)

SRC is operated by the graduate school of the University of Wisconsin-Madison and is funded entirely by the National Science Foundation (NSF) as a user facility. At present, SRC consists of two storage rings: Tantalus I, a 240-MeV ring in operation since 1968, and Aladdin, a 1-GeV ring in the early stages of operation and commissioning. Aladdin is a high-brightness source for both UV and soft-x-ray research in the range 6-3000 eV. It has 36 primary ports, each averaging 50 milliradians. At present, 22 of these ports are being equipped with beam lines for synchrotron research, 9 by PRTs and 13 by the SRC either with new monochromators specially designed for high brightness or with instruments currently installed on Tantalus but with upgraded beam-line optics to take advantage of the high brightness capability of Aladdin. Four more PRTs have been proposed but are not yet funded. This represents about 100-150 scientists from all parts of the country. Besides the normal bending-magnet sources, there are four long straight sections, three of which are available for installation of wigglers, undulators, and free-electron lasers. One bending magnet port is dedicated to use as an inverse-Compton-scattering source of gamma-ray photons at energies of ~50 MeV and greater.

Synchrotron Ultraviolet Radiation Facility (SURF)

SURF II is a 280-MeV storage ring at the National Bureau of Standards (NBS) and is dedicated to production of synchrotron radiation. The small beam cross section results in a high-brightness source for photon energies up to 400 eV. There are 11 experimental stations nearly operational, which are used by about 50 scientists mainly from NBS and the Naval Research Laboratory (NRL), together with their collaborators. In addition there are PRTs from the NRL and the University of Maryland. An active area of research at SURF II is atomic and molecular gas-phase spectroscopy, including the spectroscopy of laser-excited states.

Research Highlights of the Past Decade

Research with synchrotron radiation has evolved tremendously during the last decade or so, as indicated by the fact that only one of the five national facilities existed as a dedicated storage ring in the early 1970s. Almost all of the many ways in which photons have been put to use in condensed-matter physics have been revitalized by synchrotron radiation (SR) studies. This is illustrated in Table E.2, which summarizes the many techniques in use in SR studies in the vacuum ultraviolet (VUV) and x-ray spectral ranges.

Electronic structure. A new era in experimental studies of solid-state electronic structure began in the 1970s as emphasis shifted from the use of more conventional absorption and reflection studies to the use of photoelectron spectroscopy. Angle-resolved photoelectron spectroscopy (ARPES) coupled with SR tunability has been used to determine experimentally the energy-momentum dispersion, $E(\mathbf{k})$, for elements such as Cu and Ni, as well as for more complex materials such as GaAs and CdS. This allows the calculated $E(\mathbf{k})$ relation to be compared directly with experimental data rather than indirectly through calculated optical constants.

Magnetism. Typical of the studies possible with SR on magnetic materials are results from photoemission spectroscopy of nickel. The d-band width is experimentally determined to be ~25 percent smaller than the value predicted by the best theoretical band calculations. The measured exchange splitting near the Fermi energy is ~50-60 percent of its calculated values. The temperature dependence of the exchange splitting has been measured through the Curie temperature and demonstrates the inadequacy of a purely band model of ferromagnetism.

Structural Studies and Phase Transformations. SR studies have contributed to our understanding of phase transitions during the past decade. One example is two-dimensional (2-D) melting and wetting.

TABLE E.2 Techniques in Use in Synchrotron Radiation Studies in the Vacuum Ultraviolet (VUV) and X-Ray Spectral Ranges[a]

Measurement	Tunability	Brightness	Pulsed Time Structure	Polarization	Total Intensity	Remarks
VUV						
Absorption	●	◐		◐	○	Optical constants of
Reflection	●	◐		◐	○	materials
Ellipsometry	◐	●		●	◐	
Luminescence	○	◐	●	◐	◐	Exciton dynamics
Fluorescence	●	◐	○	○	◐	Molecular solids
Photoelectrons						
integrated	◐	◐		◐	◐	First SR experiments, 1972
angle-resolved	●	●	●	◐	◐	First SR experiments, 1975
surfaces	◐	●	○	○	◐	Surface shifts observed, 1980
Photoionization	◐	○			◐	
Photodesorption	○	◐	◐	○	◐	
Laser-induced absorption	○	●	◐	◐	◐	
X ray						
EXAFS	●	◐		○	○	First SR experiments, 1974
SEXAFS	●	◐		◐	●	Routine use requires wiggler
XPS	○	◐		◐	◐	surface structures
Raman	◐	○	○	○	◐	
Holography	◐	○		○	◐	
Fluorescence	○	◐		○	●	
Topography	◐	●	○	○	◐	
Scattering	○	●		◐	○	High-Q resolution
Anomalous scattering	●	●		◐	○	Phase problem
Microscopy	●	◐	○	○	◐	
Lithography	○	◐			●	
Angiography	◐	●	○		◐	32-keV requires wiggler

[a] ●, Very important; ◐, moderate importance; ○, some importance.

Scattering studies at various wavelengths have been used on glassy amorphous materials to probe selectively the radial distribution functions of Ge, Se, and As atoms. In addition, the high resolution of SR in a small-angle x-ray scattering (SAXS) geometry has been demonstrated, together with its versatility in studies of phase transitions, particularly in macromolecular systems such as liquid crystals.

EXAFS and NEXAFS. The first SR studies of extended x-ray absorption fine structure (EXAFS) were performed in 1974 by monitoring the transmission of x rays through a gas sample. Since then a large number of secondary detection techniques have been developed: fluorescence, Auger, photon desorption, and total yield, whose spectra contain the EXAFS information and can be used for studying liquid and solid electrolytes in greater dilution, surfaces, complex alloys, and amorphous systems. One cannot summarize adequately here the advances due to EXAFS in condensed-matter studies of ionic conductors, metallic compound formation, mixed-valent, layered, amorphous, and impurity systems. A typical example from the field of catalysis is the EXAFS study of the bimetallic, supported Pt-Ir catalyst, which shows different local structure around Pt and Ir atoms during catalytic reactions. Studies of near-edge x-ray absorption fine structure (NEXAFS) began in about 1979-1980 as a separate area of study with SR, since it was recognized that one could extend band theory into this range and achieve an understanding of the sharp, complex peaks at energies of 0-50 eV, below the EXAFS range. Edge structure, together with data on model compounds, has been used to determine the oxidation state and local geometries of molecules on surfaces.

Surface Structure. Along with phase transition studies of surfaces and other 2-D systems have been rapidly growing areas of research owing in part to SR techniques. The development of the surface-extended x-ray absorption fine structure (SEXAFS) has been a major accomplishment of SR, providing much more accurate bond lengths than those obtained with older methods. The first results, on metal surfaces, were published in 1978, and by 1981 workers had extended the use of Auger detection to semiconductors and molecular solids and had developed total yield and ion desorption modes of detection as well. The high brightness of a wiggler or undulator beam line has been used to extend SEXAFS studies to a concentration of ~1/10 monolayer. Complementary to the local structure provided by SEXAFS are studies of surface x-ray scattering, which helped to clarify the nature of surface reconstruction on Ge(100) 2 × 1 surfaces and demonstrated a commensurate-incommensurate transition in the melting of lead on the (110) face of Cu.

Interface Studies. There have been numerous SR studies on the formation of semiconductor-oxide, semiconductor-semiconductor, and semiconductor-metal interfaces, motivated in part by the widespread technological importance of these systems. Some of the notable results are detection of core-level chemical shifts for atoms that are one to three layers from the interface, modification of Schottky-barrier height that is due to cooperative interdiffusion processes, and determination of heterojunction band-gap discontinuities by photoemission. In the past 3 years several groups have added the capability for *in situ* preparation of interface samples by molecular-beam epitaxy (MBE) which gives layer-by-layer control over the samples.

Topography. Compared with those of conventional sources, SR has made spectacular improvements in the field of topography; the SR apparatus is simpler and cheaper. Stress propagation studies have been carried out under dynamic conditions with the use of a video imaging detection scheme.

Lithography. Research in x-ray lithography is in progress at a number of industrial laboratories, in which conventional sources are used; nonetheless, there is considerable interest in the use of SR for this purpose because its intensity is 10^2-10^3 times greater than that of laboratory sources. Structures as small as 7 nm have been produced, but most of the recent work is directed toward structures with high aspect ratios, e.g., 3-4 μm widths with 0.02-0.1 μm vertical edge orientations. An example of such a structure is shown in Figure E.1. The high brightness of SR allows one to use relatively simple and well-defined processing procedures in spite of their low intrinsic sensitivity.

Microscopy. X-ray microscopy with SR is still in its infancy, but initial results look promising. Scanning x-ray microscopes with elemental selectivity were first demonstrated in 1973-1974 but with only 1-2 μm resolution. Recent studies using soft x rays (1.5-4.5 nm) have increased this resolution by an order of magnitude. Although many biological applications exist for x-ray microscopy there are important technological applications as well, since the specimens can be studied in high-pressure or liquid environments.

Time-Resolved Studies. The high brightness and intensity of SR has great potential for allowing real-time dynamical processes to be followed, but comparatively few such studies have been completed. Most common are fluorescence lifetime studies that use the repetitive pulsed nature of the SR beam. Time-of-flight (TOF) spectrometers are

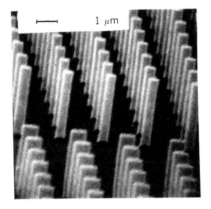
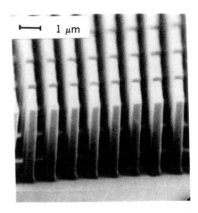

FIGURE E.1   An example of a high-aspect-ratio microscopic structure formed by etching a photolithographic pattern made by exposure to 10 Å synchrotron x rays. Note in particular the diffraction-limited straightness of the vertical walls (<0.1 μm). (Courtesy of IBM Thomas J. Watson Research Center.)

now commonly used for the photoelectron spectroscopy of gases and for ion desorption studies of surfaces. Time-resolved diffraction was first demonstrated on biological samples in 1976. Recently the structural changes of silicon during laser annealing studies were observed with a time resolution of 10 ns, and time-resolved EXAFS studies have probed local structural changes at the 10-μs to 10-ms level. Such experiments usually require single-bunch operation of the storage ring, and special time periods are assigned to time-resolved studies. New advances in detector schemes are likely to make time-dependent studies more frequent and productive in the future.

Future Directions in Synchrotron Research

In a rapidly expanding field such as synchrotron radiation research it is extremely difficult to predict what new exciting advances will occur. However, many of the recent developments mentioned above have merely demonstrated the feasibility of a new measurement or instrument. The past has shown that with each increase in resolution or intensity one has the opportunity for new areas of understanding.

EXAFS. With new and/or planned wiggler beam lines the concentration limits for EXAFS of 1 part in 10^3-10^4 will be pushed to 1 part in 10^5-10^6 routinely. This is extremely important for condensed-matter studies since many interesting defect properties occur in the 10^{-6} concentration range. SEXAFS will become a common structural tool used in conjunction with other surface probes.

Scattering. High-resolution elastic scattering will be applied to smaller samples, such as quasi-one-dimensional conductors and other novel materials. In addition, high-resolution 3-10 meV inelastic x-ray scattering will be developed to complement both SR studies of phase transitions and elastic- and inelastic-scattering studies with neutrons. Surface scattering will become a routine tool for surface crystallography and will be used in concert with other structural probes such as SEXAFS. Nuclear resonance scattering promises x-ray beams with unprecedented spectral purity.

Photoelectron Spectroscopy. High resolution in both energy and momentum will be achieved with $\Delta E < 50$ meV and $\Delta k < 0.01$ Å^{-1}. This will allow more accurate experimental determination of electronic structure, as well as studies of systems with small Brillouin zones, such as reconstructed semiconductors and novel superlattice materials. Measurements with detection of photoelectron spin will become routine, in spite of the 10^{-14} loss of intensity for spin-resolved compared with conventional measurements.

Applied Research. Although many fundamental studies are motivated by the technological applications of new materials, SR experiments have usually not been coupled strongly to high-technology materials preparation. This is already changing rapidly with advances such as the development of on-line MBE and catalytic reactor systems. Future development of realistic processing capability in lithography and microscopy has the potential of involving SR more directly in

TABLE E.3 Principal National Thermal Neutron User Facilities Operating in the 1980s

Facility, Location	Power	Commissioning Year	Average Thermal Flux (10^{14} n/cm s)	Peak Thermal Flux (10^{14} n/cm s)	Number of Instruments	Mode of Operation
High Flux Isotope Reactor (HFIR), Oak Ridge National Laboratory	100 MW[a]	1966	13	—	9	Quasi-continuous
High Flux Beam Reactor (HFBR), Brookhaven National Laboratory	60 MW[a]	1965	10	—	11	Quasi-continuous
National Bureau of Standards Reactor (NBSR), Gaithersburg, MD	10 MW[a] 20 MW[b]	1969 1983[b]	1 (2)[b]	—	9	Quasi-continuous
Missouri University Research Reactor (MURR), Columbia, MO	10 MW[a]	1968	1	—	8	Quasi-continuous
Intense Pulsed Neutron Source (IPNS)	500 MeV[a] 15 μA	1981	0.002	4[c]	7	6 mo/yr
Los Alamos Neutron Scattering Center Los Alamos National Laboratory	800 MeV[a] ~5 μA (~100 μA)[d]	1978 (1986)[d]	0.001 (0.02)[d]	0.5[c] (100)[c]	3 (8)[d]	6 mo/yr (10 mo/yr?)

[a] Thermal power for reactors, accelerator characteristics for spallation sources.
[b] Anticipated power increase.
[c] In many applications this figure somewhat overestimates performance relative to steady-state sources.
[d] After upgrade by addition of proton storage ring (PSR).

applied research. There will be a more rapid introduction of advanced materials in SR instrumentation through these applied studies. This is particularly important in developing the new high-resolution instruments to be used with wigglers and undulators.

NEUTRON-SCATTERING FACILITIES

Description of Existing U.S. Facilities

Table E.3 lists the principal user-oriented thermal neutron-scattering facilities in the United States, together with some of their important operational parameters. There are six other research reactors currently operating in the United States. One of these, the Oak Ridge Research Reactor (ORR), is a 30-MW reactor constructed primarily for engineering studies but at which three neutron-scattering instruments are in productive use by the Ames Laboratory staff. (The Ames research reactor was shut down in 1977.) The remaining reactors, at the Massachusetts Institute of Technology (MIT), the University of Rhode Island, the Georgia Institute of Technology, and the University of Michigan, while they have smaller thermal fluxes than those listed in Table E.1, continue to make useful contributions not only for training students but, when used with imagination, for carrying out fundamental studies of exploratory nature, for example in such areas as neutron interferometry.

As noted in Table E.3, comparison of neutron sources is complex. It depends on the configuration and size of the source and beam tubes (which affect intensity at the sample and fast neutron contamination), the design of the instruments, and other parameters. With pulsed (spallation) sources, neutron burst times and pulse repetition rates are also important in various degrees, depending on the experiments under consideration. Intercomparison of steady-state and pulsed sources is more complicated yet. Over the years reactors have proven to be stable, reliable sources of thermal neutrons, well suited to straightforward diffraction measurement of incident and scattered-neutron energies. Such instruments have been, and probably will continue to be, the instruments of choice for most elastic- and inelastic-scattering measurements on single crystals, traditionally the largest sector of neutron-scattering research. In conjunction with cold-neutron moderators, reactors are also well suited for the production of low-energy neutrons, desirable for high-energy resolution and small-angle scattering studies.

However, spallation sources have useful and interesting characteristics:

• The neutrons are produced in short (~10-μs) repetitive bursts. Peak intensities ~100 times larger than the continuous flux of the best reactor sources are possible with moderate average target power densities.

• The neutron spectrum can be tailored to produce relatively large numbers of higher-than-thermal-energy neutrons, so-called epithermal neutrons, with energies of about 1 eV.

• The pulsed nature of a spallation source can be used to advantage in combination with time-of-flight energy-measuring techniques that are well suited to measurements requiring information over a wide range of energy and/or momentum. This technique is particularly well suited to powder diffraction, as the resolution can be made essentially constant over a wide range of momentum transfer.

Specific instances where these characteristics can be put to use are considered in a later section.

Research Highlights of the 1970s

Neutrons have made incisive contributions to many areas of condensed-matter physics in the last decade as the chronology of Table E.4 demonstrates. These contributions are explicit in many of the preceding subsections of this report. Let us briefly review them here.

Electronic Properties.   Neutron scattering has proven to be the best method for probing low-lying electronic energy levels, for example, the localized f-electron levels in rare-earth and actinide metals. Neutron scattering was instrumental in elucidating the behavior of charge-density waves (CDWs) in quasi-1-D and 2-D metals. In particular, the phenomenon of CDW lock-in, by which the wavelength of the CDW, initially determined by the geometry of the Fermi surface, is forced into registry with the underlying lattice, emerged from a neutron-scattering study of $TaSe_2$.

Phonons and the Electron-Phonon Interaction.   While optical spectroscopy provides detailed information about long-wavelength phonons, and electron tunneling is valuable in particular cases, inelastic neutron spectroscopy provides the only systematic way to investigate the complete phonon spectrum of bulk condensed matter. These results are used directly to deduce the nature and strength of interatomic forces. Among the many noteworthy achievements of the last decade we must include the following: studies of the electron-

TABLE E.4 Summary Highlights of Neutron Studies of Condensed Matter

Date	Subject	Nation
1970-1973	Dynamics of structural phase transformations	U.S., Canada
1970-1974	Studies of low-dimensional magnetic systems	U.S.
1972	Dynamics of classical liquids	U.S.
1973	Dynamics of quantum solids	U.S.
1973-1974	Diffusion of H in metals	Germany, U.S.
1970-1974	Development of neutron guide tubes	Germany
1974	Development of cold-neutron moderators	France
1975	Powder-profile crystallography	Netherlands, France, U.S.
1975-1978	Development of microvolt spectroscopy (backscattering and spin echo)	Germany, Hungary, France
1976	Charge-density waves and incommensurate systems	France, U.S.
1976	Conformation of polymers	France, Germany
1976-1977	Development of small-angle-scattering facilities	Germany, France, U.S.
1976-1979	Studies of adsorbed monolayer films	U.S.
1976	Production and storage of ultracold neutrons	Germany, France
1976-1977	Neutron interferometry	U.S., Austria
1978	Dynamics of liquid ^{3}He	U.S., France
1979 to date	Magnetic order in superconductors	U.S.
1980	Dynamics of polymers and turbulent flow in liquid crystals	France, Norway
1980 to date	Spallation neutron sources	U.S., Great Britain, Japan

phonon interaction in metals, particularly of transition-metal compounds, which stimulated tractable first principles calculations; observation of giant phonon frequency anomalies (Kohn anomalies) in quasi-1-D metals; fundamental studies of the interplay between phonons and superconductivity; the observation of additional acoustic-like excitations—phasons—in incommensurate structures.

Magnetism. There is no area of investigation in condensed-matter physics more inextricably coupled to neutron scattering than magnetism. The determination of magnetic structures was one of the first applications of neutron scattering, and this technique will continue to be essential for this purpose into the indefinite future. One of the most active areas of inelastic neutron scattering in the 1970s concerned the nature of spin excitations in various magnetic media. It was found, for example, in suitably anisotropic materials, that 1-D and 2-D spin waves continue to propagate freely above the magnetic ordering temperature. More highly localized nonlinear disturbances, known in classical mechanics as solitary waves, have been shown to have quantum-mechanical analogs in certain 1-D magnets. Inelastic-scattering measurements have also probed the nature of spin excitations in amorphous magnets and provided relaxation times for magnetic disturbances in both valence fluctuation and spin-glass systems.

Critical Phenomena and Phase Transformations. The 1970s may well be remembered among future condensed-matter physicists as the decade of the phase transformation, and neutron-scattering experiments on magnetic systems played a leading role in contributing to this perception. Of the experimental quantities of interest (see Chapter 3)—correlation length, order parameter, susceptibility, and heat capacity—neutrons alone provide measurements of the first and, in the case of antiferromagnetism, of the second of these quantities as well. The decade began with the demonstration of large critical magnetic fluctuations in the ferromagnetic metals iron and nickel, progressed through careful quantitative neutron studies of critical exponents in antiferromagnets, the effects of dimensionality in quasi-1-D and 2-D magnetic systems (including experimental verification of Onsager's famous solution of the 2-D Ising model), and the effect of long-ranged dipolar interactions, and continues most actively at present in studies of the effect of random fields and impurities. Important neutron-scattering studies were also performed on structural phase transformations, particularly with regard to the dynamical role of unstable phonon modes as active agents in driving structural transformations.

Surfaces. Pioneering studies of the structures of physisorbed layers of simple gases at coverages ranging from less than one monolayer to several layers thickness have been carried out by neutrons on bulk samples of graphite that have been exploded in a controlled way to produce very large numbers of internal, but essentially free, surfaces upon which materials can be adsorbed. Depending on the conditions of temperature and pressure, several sorts of adsorbate structures are observed: commensurate crystalline phases in which the interatomic separations of the adsorbed layer accommodate to the underlying graphite lattice, floating or incommensurate films that are unregistered with the graphite, and 2-D fluids without well-defined long-range order. Molecular gases may be freely rotating or precisely oriented with respect to the substrate. Some of these effects are quite subtle and benefit from the higher-resolution studies that are now possible using synchroton x-ray sources. It is important to remark that delicate questions concerning the nature of long-range order in two dimensions, currently of great fundamental interest, can only be addressed by diffraction measurements. Thus, the continuing value of x rays and neutrons to surface studies is assured despite the spectacular recent successes in direct imaging of surface structure using electron microscopes. Neutrons have also begun to contribute to inelastic-scattering studies of physisorbed and chemisorbed surface phases.

Polymers. The fact that the 1970s witnessed major progress and growth in application of small-angle neutron scattering to polymer physics is directly traceable to the technique of isotopic contrast manipulation, which in essence works as follows. By favorable circumstance hydrogen and deuterium differ greatly in their neutron-scattering power. Thus a hydrogen-containing macromolecule, or a given segment of it, can be labeled by selective deuteration, with the result that neutrons sense only the presence of these labeled molecules when mixed in a dilute solution of their chemically similar undeuterated counterparts (see Figure E.2). The problem waiting for this tool was the polymer conformation question, i.e., what is the shape assumed by an individual macromolecule in a polymer melt? The answer, settling a long-standing theoretical controversy, is that the molecules assume the shape of a random coil, a behavior paradoxically more ideal than that of polymers in dilute solution, for which the coils are swollen by interaction effects. Similar conformation studies have now been carried out on quenched polymer melts, as well as crystalline and semicrystalline solids and stretched polymers, thereby addressing fundamental questions in polymer rheology. Micro-electron-volt inelastic neutron scattering has just begun to examine the dynamics of polymer segments at previously unexplored length scales. Although it lies outside the scope of this report, it must be stressed that the use of H/D contrast variation has enormously enhanced the power of neutron scattering in structural studies of biologically functional macromolecules as well.

New Materials. Condensed-matter physics is continuously renewed by the creation of novel materials, and neutron scattering plays an essential role in characterizing the magnetic and dynamical properties of these new materials. A list of recent examples of such materials in which neutron studies figure prominently includes 1-D metals, graphite intercalation compounds, and magnetic multilayer films. Another case in which the unique capabilities of neutron measurements are readily apparent concerns magnetic superconductors, more pre-

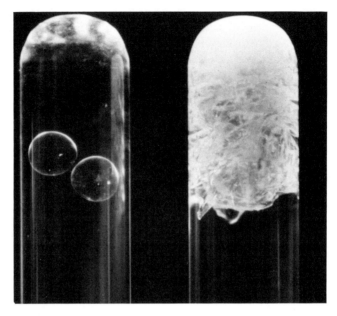

FIGURE E.2   Both tubes contain two Pyrex beads imbedded in glass wool. They become visible in the tube on the left when filled with a solvent with a refractive index matching that of the glass wool (slightly smaller than that of Pyrex). In a similar way individual segments of complex macromolecules are rendered visible to neutrons by adjusting the neutron-scattering power by selective deuteration.

cisely the coexistence of magnetic and superconducting long-range order, which was discovered (engineered is a better term) in the late 1970s. For although in these new materials heat-capacity measurements indicated the presence of some kind of ordering occurring below the onset of superconductivity, elastic neutron scattering was necessary to establish its nature, distinguishing between antiferromagnetism, which can coexist with superconductivity, and ferromagnetism, which suppresses it. The competition between superconductivity and ferromagnetism is subtle and interesting, as revealed by the appearance in small-angle neutron-scattering experiments of a new type of long-wave-length oscillating magnetic disturbance present at temperatures above those at which ferromagnetism displaces superconductivity (Chapter 8).

Future Directions

As is the case in all experimental disciplines, new opportunities are closely coupled with technical advances in instrumentation. In what follows, we identify some of these advances as well as the attendant new physics that they can be expected to address.

Epithermal Neutrons

Increased epithermal neutron fluxes at spallation sources will make possible further new classes of experiments at unexplored regions of energy and momentum. Among these new experiments are the following:

- The study of high-frequency vibrations of a solid, particularly those involving hydrogen atoms. Overtone and combination modes of excitation can be studied to deduce the shape of the vibrational potential energy surface.
- The study of magnetic excitations in solids is generally restricted to small momentum values. Kinematic restrictions in the scattering make it necessary to measure with 1-2 eV epithermal neutrons even for rather moderate (~ 100-meV) magnetic excitation energies. The nature of magnetic spin waves and the continuum of single-particle excitations into which they dissolve is an important example of a long-standing issue that can be addressed.
- In the high-momentum transfer limit scattering events in condensed matter can be considered as taking place from individual atoms. These deep inelastic-scattering events allow a direct measurement of the momentum distribution of light atoms in their ground state. These measurements are particularly important for superfluid liquid ^{4}He, for which there presumably exists a zero-momentum condensate fraction.
- The larger accessible range of momentum transfer provides, through Fourier transformation, increased spatial resolution in structural studies of liquids and amorphous solids.

Neutron Guides

In modern thermal neutron sources, the high flux is of necessity generated in a small (~ 1 m^3 or smaller) core or target region. There are obvious practical limitations to the number of instruments that can be tightly grouped around the high-flux region. An innovation that occurred too late to be used to optimum advantage in the U.S. high-flux reactors is the neutron guide, which uses neutron mirrors to transport a beam of neutrons large distances from core to instruments with negligible loss. Because neutron mirrors collect over a solid angle proportional to the square of the neutron wavelength, existing guide tubes work best for long-wavelength cold neutrons. Supermirrors, conceptually related to multilayer optical antireflection coatings, promise to extend the use of neutron guides to shorter wavelengths. These developments allow additional instruments to be added to existing facilities once considered saturated.

High-Resolution Spectroscopy

Typically, the energy resolution of conventional thermal neutron spectrometers is of the order of 0.1 meV or more. While this is usually adequate to determine the mean energy of excitations in condensed matter, there is much information contained in the widths and lineshapes, particularly in solids at low temperature, which are not accessible at this level. Unconventional instruments using, for example, neutron spin precession to encode the energy information of the scattering event, and having energy resolution of 1 μeV or less, are now in routine use in Europe. They have opened up new areas such as molecular tunneling spectroscopy in molecular crystals and the investigation of low-frequency polymer dynamics. No micro-electron-volt resolution instruments currently exist in the United States.

Interferometry

The advent of monolithic single-crystal Si interferometers has made it possible to measure neutron wavelengths to unprecedented accuracy. The technique has thus far been used to measure gravitational and quantum-mechanical properties of the neutron. In the coming years, we may expect to see interferometry turn increasingly to questions of interaction and propagation of neutrons in condensed matter.

Growth of the Neutron User Community

Two important developments occurred in the 1970s that have begun to restructure the neutron community, attracting appreciable numbers of part-time neutron scatterers from universities and industry:

• Powder profile analysis is a technique involving extensive computer analysis of powder diffraction patterns to extract incompletely resolved features. This permits the determination of crystal structures four to five times more complex than was previously possible. Since the measurements themselves are rather fast and routine, the technique has become popular among chemists and crystallographers without previous neutron-scattering experience.

• Small-angle neutron scattering (SANS) is another technique that is well suited to active outside user participation in the actual data-taking stage. The powerful contrast matching techniques described previously in this study have caused an unparalleled growth of interest in neutron scattering among biologists, polymer scientists, and materials scientists.

Principally as a result of these developments and the clarification of DOE policy toward outside users, which occurred in 1979, the U.S. neutron-user community has embarked on a course of rapid expansion in recent years, growing from about 250 per year in 1977 to over 500 per year in 1983. It will be important in the coming decade to continue to nurture these two areas of active user involvement. It is also likely that the spreading use of synchrotron radiation sources will trigger an increased use of neutrons. As more scientists break from the traditional mold, restructuring their research programs around distant synchrotron facilities, they may simultaneously become more aware of the complementary nature of neutron probes and the strong similarities in methodology and instrumentation that underlie x-ray and neutron-scattering experiments.

HIGH-MAGNETIC-FIELDS FACILITIES

Description of U.S. Facilities

The National Magnet Laboratory (NML) at MIT is the only major user facility for high field research and development in the United States. It has a scientific and technical staff of about 60 and an operating budget of $10 million/year, of which approximately one half is the core support from the NSF for operations, magnet development, and in-house research.

The NML facility has a wide variety (24) of steady field magnets with different peak fields, bore size, and homogeneity. The highest field magnet is a hybrid (superconducting solenoid plus conventional magnet) configuration, which holds the world's record for dc fields at 30.4 T (1 tesla = 10^4 oersteds). In the past year, a 45-T pulsed field facility has become available to users at NML, and engineering studies are under way for the development of 50-70 T pulsed configurations.

Research Highlights of the 1970s

Much of the in-house effort at NML during the past decade has been dedicated to the development of higher field magnets, through the better understanding of superconducting materials, and to their utilization in both research and applications. Notable among these efforts have been the measurements of the upper critical field H_{c2} in the Al5 (e.g., Nb_3Sn) and Chevrel phase superconductors, the development of the spin-polarized tunneling technique for probing spin-orbit scattering in high field superconductors and the spin polarization in magnetic materials, high-gradient magnetic-field separation (now used commercially to purify clay), the development of magnetoencephalography, and the development of ductile, high-current superconducting wire for high-field magnets. Extensive work has been carried out at NML on magnetic phase diagrams; the observation of the Lifshitz point in MnP was one important accomplishment in this in-house NML program. A major effort in quantum-optics studies in semiconductors (e.g., spin-flop Raman scattering in CdSe) has been in place at NML for more than a decade.

Perhaps the most exciting research requiring high fields that has taken place at NML by outside users has been the probing of the electron dynamics of 1-D and 2-D systems. This has led to the discovery of the quantization of the Hall effect and, most recently, of fractional integer filling of the lowest Landau level in GaAs/(Ga,Al)As layers. It is now believed that the ground state of the 2-D electron gas in a high magnetic field is a highly correlated liquid rather than a crystal. Of considerable interest are the magnetotransport experiments on graphite, which show that large (25-T) fields induce a charge-density wave.

Other notable work utilizing high magnetic fields has been the studies of the spin-Peierls transition—an intrinsic lattice instability in a 1-D, $S = 1/2$ antiferromagnetic chain coupled to a 3-D lattice; the properties of itinerant ferromagnetic and antiferromagnetic metals; spin fluctuation (e.g., UAl_2) and valence fluctuation (e.g., YCuAl) systems; and electron spin resonance in the submillimeter region.

Opportunities

There are a variety of novel and fascinating phenomena that occur in fields that are already available or that will become available in the next decade, with the advent of pulsed fields in the 70-150 T range and beyond. (Nondestructive fields as large as 150 T of microsecond duration have recently been achieved in Japan.) Among the possibilities is the observation of superconductivity based on p-wave pairing, which occurs in liquid ^3He but has never been found in an electronic system. Strong magnetic fields should also make possible the orientational ordering of large molecules via their diamagnetic susceptibilities, measurements of the nonlinear magnetic response of transition and actinide metals, and exploration of the magnetic phase diagrams in materials whose exchange coupling is too large to be affected by the currently available fields. Magnetic fields available in the 1980s should enable us to reach the Paschen-Back limit of strong magnetic fields in hydrogenlike systems in certain semiconductors, such as excitons in InSb.

Some of the most interesting developments may be expected in conducting systems of lower dimensionality, such as semiconductor inversion layers (e.g., the common field-effect transistor), and highly anisotropic layered systems such as graphite and certain organic conductors. Already, new phase transitions have been observed in the layered systems, produced by strong magnetic fields. The experimental discovery of the quantized Hall effect serves as a dramatic demonstration that there may remain many exciting phenomena yet to be discovered in the area of condensed-matter systems in high magnetic fields.

FACILITIES FOR ELECTRON MICROSCOPY

Introduction

Electron microscopy is one of the most important multipurpose techniques in the sciences of solid-state and biological materials. Several essential modes of operation are unique to the electron probe. These include microchemistry on a scale of ≤ 10 nm by energy loss and x-ray detection; atomic-resolution imaging by lattice fringe methods and by scanning with a probe focused to a few angstroms; and convergent-beam diffraction analysis of structures 10 nm in size. In addition, microscopes can be converted to scattering instruments in which the dispersion relations of crystal excitations are examined through events that combine energy loss and momentum transfer. An important asset is that these experiments can be performed on submicrometer-size samples. Electron microscopy is one of the essential capabilities required to complement new materials development (see Appendix C), particularly in areas of microfabrication. Other modern directions include ultrahigh-vacuum instruments designed for surface spectroscopy (Figure E.3), and high-level computerization of microscopy applied to microchemistry and to diffraction from complex systems, including biomolecular assemblies.

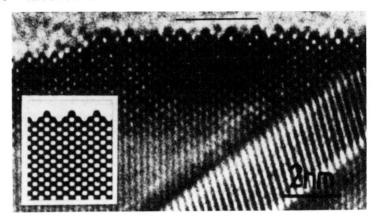

FIGURE E.3   The atoms at the surface of a crystal often reconstruct, assuming a different configuration from that of the bulk. This figure shows an electron microscope image of the profile of the surface of a gold particle showing a reconstructed surface superstructure. The inset shows a simulation corresponding to a missing-row model, which matches the experimental image. [After L. D. Marks and D. J. Smith, Nature *303*, 316 (1983).]

Description of U.S. Facilities

Most electron microscopes in the United States are dedicated to the research of individuals or small groups. Nevertheless, there now exists a wide variety of instruments employed in a user facility mode. Many campuses have electron microscopes operated by individuals but accessible to other on-site users. User operation on a national scale first started in the 1960s with several 1-MeV microscopes located at national laboratories and other public institutions. These high-voltage instruments are specially valuable in biological applications owing to their improved penetration, because they accommodate thick specimens and often possess stereo techniques for depth perception. In materials research the emphasis with high-voltage machines has been on environmental effects, on *in situ* deformation measurements, and on radiation damage produced either by the electron beam itself or by an accelerated beam of ions. The considerable expense of the machines made wide accessibility appear desirable and prompted the facility mode of operation.

The DOE supports two national user facilities in advanced electron microscopy. The National Center for Electron Microscopy at the Lawrence Berkeley Laboratory includes a 1.5-MeV high-voltage electron microscope and an atomic resolution microscope. The Argonne Center for Electron Microscopy includes a 1.2-MeV high-voltage electron microscope to which is coupled a 2-MeV ion accelerator for *in situ* radiation damage studies. A state-of-the-art analytical electron microscope at 300 to 400 keV voltage will be added in 1985. Both facilities are available to the user community at no charge, unless proprietary work is done. The DOE also supports two other user programs where advanced microanalytical electron microscopy can be performed. These are the Center for Microanalysis of Materials at the University of Illinois and the Shared Research Equipment (SHARE) Program at the Oak Ridge National Laboratory. The only other U.S. facility for high-resolution research is at Arizona State University. It operates with commercial instruments manufactured in Japan, The Netherlands, and the United Kingdom, which have been locally developed and modified for specific types of high-resolution application. Important work on basic microscope development has been carried out at the University of Chicago, where atomic-resolution scanning transmission machines have been constructed (as prototypes, not facilities). Active efforts in several areas of development and accessory construction are also in progress at other U.S. institutions.

Unfortunately, the necessary capabilities are not usually combined in a single machine. Nor are all the techniques learned quickly. For this reason there remains a need for regional centers where a number of complementary machines can be maintained, developed, and used by experts. The grouping of several experts with several machines provokes valuable interactions in a well-equipped environment and is an economically sound organization. Since regional centers of this type can serve teaching and consulting purposes for a wider community, it appears desirable that a number of centers of expertise or facilities of this type be maintained nationwide into the indefinite future. If sited in university communities, centers of this type could help to strengthen U.S. science in the area of electron microscopy both by expert ongoing research programs and through expert training of graduate students to staff microscopy efforts at other institutions nationwide.

Advances of the Past Decade in Electron Microscopy

A number of significant advances have taken place in the field of electron microscopy over approximately the past decade. These include the following:

1. The development of convenient scanning transmission electron microscopes (STEM) that yield information at atomic resolution by rastering a finely focused beam over the sample.

2. The incorporation of energy-dispersive x-ray (EDX) and electron-energy-loss spectroscopies (EELS) in transmission microscopes to make microchemical information available on a 20 Å and up lateral scale in EELS, and on a 100 Å and up scale in EDX, dependent on specimen characteristics.

3. Development of microdiffraction, including convergent beam diffraction techniques (and suitable electron microscopes), which made available routine crystal-structure identification on particles 100 Å or larger in size and definitive symmetry determination on larger particles.

4. Continued development of high-resolution machines and techniques to make most interatomic distances in solids resolvable and weak-beam high-resolution defect imaging feasible.

5. The use of electron microscopes and development of vacuum capabilities for high-resolution examination of crystal surface structure and steps, for example.

Outlook for the Future

One can expect to see important continuation of this progress over the next decade in the following areas among others:

1. Development of the electron microscope as a miniature laboratory that includes diffraction, scattering, imaging, EDX, EELS, Auger spectroscopy, cathodoluminescence, specimen current yield, and other spectroscopies directly available for analytical purposes.

2. A new generation of flexible intermediate-voltage machines (300-500 keV) with laboratory (rather than institutional) size and cost scale that may satisfy many expectations left unfulfilled by the last generation of high-voltage machines.

3. Microscopes that will be developed with ultrahigh-vacuum and sample-preparation capabilities suitable for routine exploration and analysis of clean surfaces.

Contributors to This Volume

PART I

E. ABRAHAMS, Rutgers University
P. W. ANDERSON, AT&T Bell Laboratories
A. M. CLOGSTON, Los Alamos National Laboratory
G. GAMOTA, University of Michigan, Ann Arbor
F. E. JAMERSON, General Motors Research Laboratories
D. C. SHAPERO, National Research Council
L. R. TESTARDI, National Bureau of Standards
S. WAGNER, Princeton University

Chapter 1

M. L. COHEN, University of California, Berkeley
H. EHRENREICH, Harvard University
W. KOHN, University of California, Santa Barbara
J. A. KRUMHANSL, Cornell University
D. LANGRETH, Rutgers University
J. C. PHILLIPS, AT&T Bell Laboratories
E. W. PLUMMER, University of Pennsylvania
A. L. RUOFF, Cornell University
D. SCHIFERL, Los Alamos National Laboratory
J. R. SCHRIEFFER, University of California, Santa Barbara
L. J. SHAM, University of California, San Diego

291

J. SHANER, Los Alamos National Laboratory
I. SILVERA, Harvard University
J. W. WILKINS, Cornell University

Chapter 2

B. J. ALDER, Lawrence Livermore National Laboratory
N. ASHCROFT, Cornell University
J. D. AXE, Brookhaven National Laboratory
J. L. BIRMAN, City College of the City University of New York
W. BRON, Indiana University
G. BURNS, IBM Thomas J. Watson Research Center
D. M. CEPERLEY, Lawrence Livermore National Laboratory
R. COLELLA, Purdue University
A. J. FREEMAN, Northwestern University
F. GALEENER, Xerox Palo Alto Research Center
B. N. HARMON, Iowa State University
W. C. HERRING, Stanford University
M. V. KLEIN, University of Illinois
A. A. MARADUDIN, University of California, Irvine
H. MARIS, Brown University
A. K. MCMAHON, Lawrence Livermore National Laboratory
E. J. MELE, University of Pennsylvania
D. L. MILLS, University of California, Irvine
V. NARAYANAMURTI, AT&T Bell Laboratories
R. J. NEMANICH, Xerox Palo Alto Research Center
G. E. PAKE, Xerox Corporation
G. SHIRANE, Brookhaven National Laboratory
M. F. THORPE, Michigan State University
R. M. WHITE, Xerox Corporation
J. P. WOLFE, University of Illinois, Urbana
R. ZALLEN, Virginia Polytechnic Institute and State University

Chapter 3

C. KNOBLER, University of California, Los Angeles
P. C. HOHENBERG, AT&T Bell Laboratories

Chapter 4

J. C. BONNER, University of Rhode Island
B. R. COOPER, West Virginia University
C. Y. HUANG, Los Alamos National Laboratory
I. S. JACOBS, General Electric Company
V. KORENMAN, University of Maryland

D. P. LANDAU, University of Georgia
R. E. PRANGE, University of Maryland
S. SCHULTZ, University of California, San Diego
D. J. SELLMYER, University of Nebraska

Chapter 5

M. BRODSKY, IBM Thomas J. Watson Research Center
J. DEMUTH, IBM Thomas J. Watson Research Center
A. FOWLER, IBM Thomas J. Watson Research Center
F. HIMPSEL, IBM Thomas J. Watson Research Center
N. HOLONYAK, JR., University of Illinois, Urbana
T. HO, IBM Thomas J. Watson Research Center
N. D. LANG, IBM Thomas J. Watson Research Center
K. PANDEY, IBM Thomas J. Watson Research Center
S. PANTELIDES, IBM Thomas J. Watson Research Center
K. TU, IBM Thomas J. Watson Research Center
A. WILLIAMS, IBM Thomas J. Watson Research Center

Chapter 6

J. M. COWLEY, University of Arizona, Tempe
A. V. GRANATO, University of Illinois, Urbana
G. JACUCCI, University of Trento, Italy
N. L. PETERSON, Argonne National Laboratory

Chapter 7

M. CARDILLO, AT&T Bell Laboratories
D. HAMANN, AT&T Bell Laboratories
E. W. PLUMMER, University of Pennsylvania
T. RHODIN, Cornell University
A. J. SIEVERS, Cornell University
S. Y. TONG, University of Wisconsin
J. TULLY, AT&T Bell Laboratories
M. B. WEBB, University of Wisconsin

Chapter 8

T. BERLINCOURT, Office of Naval Research
R. BUHRMAN, Cornell University
B. MAPLE, University of California, San Diego
C. VARMA, AT&T Bell Laboratories

Chapter 9

N. ASHCROFT, Cornell University
J. A. BARKER, IBM San Jose Research Laboratory

E. G. D. Cohen, Rockefeller University
R. J. Donnelly, University of Oregon
G. Evans, Oregon State University
W. Gelbart, University of California, Los Angeles
J. Kestin, Brown University
R. McPhail, University of California, Los Angeles
R. Meyer, Brandeis University
R. L. Scott, University of California, Los Angeles
G. Stell, State University of New York, Stony Brook

Chapter 10

A. Heeger, University of California, Santa Barbara
E. Helfand, AT&T Bell Laboratories
R. Stein, University of Massachusetts, Amherst

Chapter 11

G. Ahlers, University of California, Santa Barbara
R. P. Behringer, Duke University
F. H. Busse, University of California, Los Angeles
R. J. Donnelly, University of Oregon
J. D. Farmer, Los Alamos National Laboratory
J. Ford, Georgia Institute of Technology
P. C. Hohenberg, AT&T Bell Laboratories
B. A. Huberman, Xerox Corporation
J. S. Langer, University of California, Santa Barbara
A. Libchaber, University of Chicago
L. P. Kadanoff, University of Chicago
P. C. Martin, Harvard University
J. McLaughlin, Clarkson College
A. C. Newell, University of Arizona
H. L. Swinney, University of Texas, Austin
R. W. Walden, AT&T Bell Laboratories

Appendix D

R. Brewer, IBM San Jose Research Center
Y. R. Shen, University of California, Berkeley
J. J. Wynne, IBM Thomas J. Watson Research Center

Appendix E

J. M. Cowley, Arizona State University
V. Jaccarino, University of California, Santa Barbara
J. Rowe, University of Florida

Index

295

4055